Our Oldest Task

Our Oldest Task

Making Sense of Our Place in Nature

ERIC T. FREYFOGLE

The University of Chicago Press Chicago and London

The University of Chicago Press, Chicago 60637
The University of Chicago Press, Ltd., London
© 2017 by The University of Chicago
All rights reserved. No part of this book may be used or reproduced
in any manner whatsoever without written permission, except in the
case of brief quotations in critical articles and reviews. For more in-
formation, contact the University of Chicago Press, 1427 E. 60th St.,
Chicago, IL 60637.
Published 2017
Printed in the United States of America

26 25 24 23 22 21 20 19 18 17   1 2 3 4 5

ISBN-13: 978-0-226-32639-9 (cloth)
ISBN-13: 978-0-226-32642-9 (e-book)
DOI: 10.7208/chicago/9780226326429.001.0001

Library of Congress Cataloging-in-Publication Data

Names: Freyfogle, Eric T., author.
Title: Our oldest task : making sense of our place in nature /
   Eric T. Freyfogle.
Description: Chicago : The University of Chicago Press, 2017. |
   Includes bibliographical references and index.
Identifiers: LCCN 2016054304 | ISBN 9780226326399
   (cloth : alk. paper) | ISBN 9780226326429 (e-book)
Subjects: LCSH: Nature and civilization. | Human beings—Effect of
   environment on. | Nature and civilization—United States. | Human
   beings—Effect of environment on—United States.
Classification: LCC CB460 .F74 2017 | DDC 910—dc23
LC record available at https://lccn.loc.gov/2016054304

♾ This paper meets the requirements of ANSI/NISO Z39.48-1992
(Permanence of Paper).

For Jane

Contents

Introduction

This is a book about nature and culture, about our place and plight on earth and the nagging challenges we face in living on it in ways that might endure. It deals with what American conservationist Aldo Leopold once termed the "oldest task in human history," the task of living on land without degrading it.[1] By land, Leopold meant not just soils and rocks but the entire interconnected, interdependent community of life, people included. It was an ancient task, Leopold said, an essential one, and we were struggling with it much as civilizations before ours had struggled and quite often failed. Our cleverness, technology, and fecundity: all had advanced well ahead of our collective ability to align our modes of living with nature's life-giving ways.

All living creatures change the world around them simply by going about the daily business of staying alive. To change the physical world is thus inevitable and appropriate. Indeed, the community of life that we inhabit is largely the product of such changes, made by countless species going back several billion years. The challenge that Leopold saw was thus not to avoid change to nature or even to minimize it. Instead, it was to use nature in ways that kept it fertile and productive for people then living and for future human generations, if not for other species as well. Our challenge—our oldest task—was to use nature but not to abuse it.

Collectively we have had trouble with this oldest task, particularly as our numbers have risen, new technologies have emerged, and market-driven competition has overwhelmed customary restraints on land use and resource use. Laws have curtailed some of the worst practices in many countries; green technologies are gaining a bit; and green consumerism is on the rise. But overall, our trajectory has not materially changed course. We continue to alter nature in ways that seem to involve abuse rather than use and to do so on ever-larger scales—seem to, that is, but then who can really be sure given the increasingly contentious debates? How do we know when a change we've made to nature goes beyond legitimate use to become abusive?

I entered teaching over three decades ago, in a law school, where I've led courses on environmental, natural resources, and property law along with graduate readings groups on nature and culture, social justice, and conservation thought. This book arises out of this learning and instruction. It also draws together two core stands of my thinking and writing, going back as far or farther. One strand has been my effort to come to terms with our environmental problems in their full complexity, physically, socially, and morally. Land degradation is a product of human behavior and thus of the messy mix of factors and forces that motivate and shape how we act. My sense from early on was that this degradation arose proximately if not inexorably from business as usual in the modern age more than it did from individual mischief or malfeasance. We were all complicit to varying degrees, even the most well-meaning and conscientious among us. For me this recognition posed tough questions, especially about causes and responsibility. These questions gained complexity when they were examined together with our troubling, too-frequent tendencies to deny the scientific evidence of ecological ills and to resist even proven, cost-effective reforms. Plainly, the root causes of degradation run deep, among and within us.

My musings on our earthly predicament led me to wonder also whether we were being careful and thoughtful enough when we passed judgment on the physical evidence of ecological change. We were altering nature profoundly—that much was clear—and some changes seemed manifestly bad. But it didn't appear so easy many times to decide which changes to nature were acceptable or good overall and which ones instead were misguided or immoral; it didn't appear easy, that is, to distinguish between the legitimate use of nature and the abuse of it. A normative evaluation was needed to make that determination, and that evaluation, that line-drawing, required in turn some sort of measuring standard. We didn't possess such a standard—not a sound one, at least—and the work of crafting one, I sensed, was far harder than we recognized. Many factors seemed relevant to such an overall assessment or evaluation, including factors relating to social justice, future generations, and other life forms, and to the vast gaps in our scientific knowledge.

The second strand has been my broader effort simply to make sense of the place and time in which I live—my effort, in Cicero's familiar phrasing, to escape the tyranny of the present. As have many others, I have tried to step back from the modern age and to think critically about it, to identify and come to terms with the ideas, values, and sensibilities that structure how we understand the world and engage with it. This age-old task has never been a simple one. In our time it seems particularly challenging, in part due to the abundant writing flowing out of our specialized and fragmented universities and research centers, so helpful in some ways, so distracting and overwhelming in others.

What I soon recognized was that I couldn't progress far on either of these intellectual projects without make sense of the other as well. However we might assess it, our environmental plight is a central reality of our times. It offers essential evidence of how we see the world, how we understand our place in it, and how we relate to one another, other living creatures, and future generations. Similarly, we cannot grasp why we have such trouble evaluating normatively our changes to nature, or why we bicker so about alleged ills and reform options, without broadening the inquiry greatly. In complex ways our ills have much to do with the culture of our era, with the secular, rational, and liberal values and assumptions that gained dominance in the Enlightenment era of the seventeenth and eighteenth centuries and that retain such power today. In the case of the United States they are particularly linked to the timing and peculiar political context when our nation was founded and when our collective self-identify coalesced. Adding to our challenge has been the darkening shadow of the capitalist market, now working at the global scale, the market that yields its material bounty by fostering base impulses, fragmentation, and moral and intellectual confusion—which is to say by making us, in basic ways, lesser creatures.

Out of this long-continuing inquiry has come this book. It draws upon diverse disciplines and bodies of writing, particularly history (intellectual, social, environmental), ecology and evolutionary biology, economics, social and political writing generally, and various core strands of philosophy. My home discipline, law, also plays a role—most openly in a critique of private property norms—but this is not a work on environmental or property law, present or proposed. Nonetheless, a legal perspective aids a project like this one because the legal arena is a setting in which all relevant factors on an issue are brought to bear, or ought to be. Good lawmaking is inherently wide-ranging and synthetic: It borrows facts, values, experiences, and intuitions from any and all sources. This book is synthetic in just this way: it is a patchwork effort, a labor that seeks to generate—by gathering, assembling, and assessing—emergent properties, just as natural communities do when their components come together to create traits and powers not possessed by the parts in isolation.

My central thesis is that our struggle to live sensibly within the land community is pre-eminently a cultural one, not chiefly a matter of scientific knowledge, technology, or even population, though these factors are all highly relevant. Our age faces a grave cultural crisis, a crisis not just in the sense of a loss of cultural coherence—a kind of descent into endless, misguided bickering with rising anxieties and anomie— but in the stronger sense that the central elements of our world view are simply no longer working for us, intellectually, morally, or practically, particularly when it comes to our oldest task. The foundational assumptions that frame our understandings and actions seem less and less able to bring order to the facts of the day and to provide the tools we need to make collective sense of our many challenges, indeed even to admit their very existence. Our ecological ills and our inability to think clearly about them—to think clearly about our rightful place in nature—are tied to traits embedded within and among us and to the complex ways we think about and value the world, particularly our self-images as morally worthy, rational individuals, different not in degree but in kind from all other life forms. No amount of new science and technology will help us, not enough at least. No amount of green consumerism will help much either; indeed, our emphasis on "going green," I shall argue, arises out of and aggravates elements of modern culture that are themselves root causes of our planetary excesses.

This book is thus a critical inquiry into modern Western culture. It pays particular attention to the case of the United States both because I know it best and because it illustrates so vividly the cultural challenges that confront the modern age generally. The study digs into the essential components of the modern worldview, beginning with fundamental questions of reality, cognition, and morality that have long intrigued and challenged philosophers. This attention to fundamentals is essential, I believe, because our central cultural need is not, as it is sometimes said, a mere matter of learning to be nice to nature or to love "mother earth"—a shallow notion that is as confused and unfocused as it is well intended. Our needs for change go much deeper than that. Major elements of our worldview require revision.

This critical inquiry provides the foundation for a proposed recasting of our worldview, aimed at promoting better modes of living on earth and also at promoting—necessarily promoting, as I will argue—greater social justice and heightened concerns for future generations and other life forms. The inquiry pays extended attention to the challenge of distinguishing between the legitimate use and abuse of nature—a difficult challenge, one we've addressed poorly. It includes a hard look at science—what it is and is not—and at our cultural tendency to turn to science to answer questions that it simply cannot answer (even as, at the same time, we push aside scientific findings we don't like). At the same time it digs deeply into the unsteady bases of moral thinking, revealing why our search for new moral standards for the ecological age are frustrated by an overreliance on empirically grounded objectivity and by confusion over the true, social origins of the liberal values that we do embrace. Necessarily, a new, more ecological perspective on the world is needed, along with new ecological ways of recognizing how we are embedded in nature. Further, we need to think much more broadly about environmental justice and about the roles and moral status of the planet's other interconnected life forms.

Inevitably the recasting of culture undertaken in this book reflects my own value preferences as author. But the cultural framework I construct—my dissection and diagnosis of our plight—should have value also for readers of different temperament, readers perhaps less inclined than I to care intrinsically and aesthetically about other life forms or more prone to trust that future generations will be smart enough to handle whatever ecological forces we unleash.

I have written this book as I have because I know of no other book like it, no book that similarly endeavors to set the full intellectual stage. A particular hope is that it will help environmental scholars, students, and activists see how and whether their personal efforts make sense when viewed in context. Without the full picture it is hard or impossible to know. Is it sensible, for instance, to push forward the idea of ecosystem services as an intellectual frame? It is proper for biologists to treat species-preservation as a shared goal? Would a market-guided carbon-trading system, all things considered, help us come to terms with climate change? Are religion-based claims about the value of Creation inappropriate for public use? Indeed, is it simply a matter of personal opinion whether one landscape condition is better than another?

————

In its cultural malaise and confusion the modern age resembles the years during and after the American Civil War in the mid-nineteenth century. As historian Louis Menand explains, sensitive observers of that tumultuous time understood the Civil War not just as a colossal failure of democracy and goodwill but also as a failure of culture, a failure of prevailing beliefs, values, sentiments, and ideas. The war swept away the South's slave civilization and took "almost the whole intellectual culture of the North along with it."[2] This breakdown was complexly linked to the great forces then at work—industrialization, urbanization, shifts of power to impersonal bureaucracies, and more. These forces together with their widespread economic and social consequences seemed to foster cultural chaos, to fracture the moral order and derail the nation's progressive trajectory. Not until the end of the nineteenth century, Menand relates, did a new culture rise up to interpret and forge a counter to these massive forces, and it did so, in retrospect, only partially and temporarily.

The new, more confident culture of the Progressive Era ran aground not long after it emerged, when and as Western civilization descended into the disorienting horrors of World War I. Perhaps modern civilization was, after all, merely the thin veneer that doubters all along had claimed it was, a veneer that when stripped away by conflict or community fracture exposed the vast depths of inner darkness that Joseph Conrad and others had probed. For novelist Willa Cather the world broke in two around 1922; year one of the new era, Ezra Pound termed it. It was the time of *Ulysses*, *The Waste Land*, and *Babbitt*; a time, in the United States, of railroad strikes, mining massacres, and national corruption. One could, as H. L. Mencken did, ridicule the social and moral constraints of gentile and bourgeois culture, the culture that supplied the traction for reform efforts to fight corruption and decay. But where was the nation to find a new public morality that could effectively take its place and domesticate the economic and cultural forces of the day?

The cultural crisis we now experience is in many ways a continuation and strengthening of the crisis that Menand has charted and of the ensuing malaise fed on industrial-style war. To be sure, periods of peace and prosperity would come after the Great War and ensuing ones, creating times and auras of calm. But the acids of modernity had been dispersed widely. Particularly disorienting for the educated were the tales told by Darwin, Freud, and Einstein, who together cast doubt on human exceptionalism and on the objectivity and reliability of human reason and sense perceptions. Beyond that and also disorienting were the broadening demands for individual liberty and moral autonomy, the often amoral forces of industrial capitalism and bureaucratic efficiency, and the rising, dismaying evidence that the vast continent was not, after all, an unlimited warehouse of natural resources.

————

It will help at this beginning point to say more about these cultural ills, by way of introducing the inquiry to follow. As will be clear, our cultural deficiencies are intertwined, not just with one another but also with many of our cultural strengths. A quick look here at the overall picture might help clarify the pertinence of the individual parts of this wide-ranging overall inquiry, particularly the early ones.

Our cultural deficiencies fall rather roughly into four connected categories.

Moral value and interdependence. In the first category are the various ways that we commonly understand our roles and capacities in nature as the morally supreme form of planetary life and also perceive ourselves as autonomous beings. As human beings we do differ from other species of life, and they from one another. Yet we err gravely by assuming that we differ from them in kind rather than merely in degree, by assuming (implicitly today, more explicitly in the past) that we are in the long run really not much constrained by the planet's physical capacities and functioning. We profess true belief in Charles Darwin's evolutionary writings and likely chuckle at stubborn-minded contemporaries who can't seem to face evolution's bracing truths. But in an important sense those who reject a purely materialist interpretation of evolution are the more intellectually honest and coherent among us. They sense and profess that humans differ in kind from other life forms and they act consistent with this belief, as much interpretive and moral as it is factual. Evolution's defenders, in contrast, while professing their kinship with the apes nonetheless seem plenty content to hold on to the generous benefits of our presumed creaturely uniqueness.

This first category of cultural elements has to do with the locus of moral value in the universe: does it extend beyond humans to other life forms, to species, and to communities? It also draws in the composition of reality (metaphysics) and the nature of individual humans and other life forms as living beings (ontology). Is the world made up not just of physical stuff—of atoms and their components and combinations bouncing about—but also of intangibles, such as ideas, moral values (goodness), and logical relationships, which exist not just in and among human minds but outside of them, embedded in the natural order? We routinely talk, for instance, as if human rights transcended mere social convention. But do such rights really exist apart from the historical forces and particular circumstances that gave rise to their proclamation; did they exist before we recognized them and will they continue existing if we forget them? If human rights do somehow exist independently of us, then what other normative or spiritual values might similarly await our belated recognition? As for the nature of existence, are we humans best understood as autonomous creatures or are we in significant ways—more important ways, even—defined and constituted by our many roles, connections, and interdependencies?

The world's physical parts are highly intertwined, as we should certainly know. They are not merely collections of fragmented pieces and parts. When we overlook these countless interconnections it becomes too easy to ignore or downplay the essential dependence of all life on the planet's basic functional processes. It becomes easy also to overlook how the planet's millions of life forms have evolved in tandem with one another, gradually and inexorably, and how they can typically thrive only when these co-evolved interactions remain within natural bounds. To miss this fundamental truth is to miss the central role of cooperation in sustaining biotic communities. Even more it is to miss the many ways that communities of life give rise in their evolving interconnections to traits and capabilities that do not reside in the biotic and abiotic parts considered separately. In short, human-caused ecological degradation is plainly linked to what we see and don't see when we look into nature, to our (over)confidence in our ability to gain knowledge about the world, and to the human-centered ways we attribute (create) moral value within it.

Reason, science, and the origins of morality. As for the second category, we are also weighted down today—even as we have been much benefited—by what might be termed our cult of objectivity, by our tendency to exalt facts and reason and to assume that, possessed of these two tools and little more (a few liberal and utilitarian principles), we can forge public policies that are workable if not better. Eighteenth-century thinkers exalted reason and demanded factual proof as part of a mission of destruction. They sought to sweep away superstition and ignorance and to challenge traditions that lingered on with no support other than long familiarity. A particular aim of the era was to push religious rites and practices out of the public sphere and into private life, and it did so even as the era's leading intellects took for granted their embeddedness in a transcendent moral order mostly derived from Protestant Christianity.

In retrospect, the Enlightenment's intellectual tools wielded too much power as they permeated the public realm. The era's guiding insistence that all beliefs rest on facts and reason, used as a critical tool, proved forceful enough to chip away the sound along with the unsound, particularly when it came to public morality. As philosopher David Hume could see, no combination of reason and sense-derived factual information seemed able to supply a solid grounding for moral standards once the old order was fractured. With revealed religion called into question, morality needed to find a new, more sentiment-based foundation. This deficiency could remain concealed so long as the prevailing Christian moral order retained its power by inertia. But other pressures soon pushed against that customary moral order—particularly strident calls for individual liberty—and reason alone could not adequately defend it. Once disconnected from experience and inherited values, reason by itself couldn't draw a reliable line between the moral and the immoral. Morality needed to secure a new, more defensible grounding, intellectual or otherwise, especially if it was to protect against the new forms of domination and exploitation.

Many troublesome ills have come from this cult of objectivity, even as an insistence on it has also yielded vast benefits.

- There is the too-easy sense within it that we can solve problems simply by gathering more facts and using reason-based science and technology, without questioning our values or becoming better people.
- There is the related tendency to turn to science and scientists to pass authoritative judgment on allegations of grave problems when this kind of assessment inherently requires the use of normative standards—standards of goodness and morality—which science as such does not possess and cannot generate.
- There is the inability to know what to do when the facts run out, as they so often do when it comes to nature and our dealings with it. How do we deal with our enduring ignorance? Religion long provided answers, and they are still at hand.
- And there is the tendency to assume that, with morals mostly matters of personal choice, the ideal state should merely keep the peace among people, leaving individuals free to pursue their separate visions of good living (a problem much exacerbated at larger scales by clashes of cultures that do not share overarching conceptions of the good or the right).

It is telling that the Enlightenment era did not challenge and corrode all public moral values. It left alone, and indeed exalted, what Thomas Jefferson famously called the "self-evident truths" of individual rights. This embrace of individual rights, soon to spread widely, was intellectually curious given that claims of rights were, in reality, no better grounded in facts and reason than any other moral precepts (senses of duty, good character, and virtue). Jefferson may have said as much when he termed them "self-evident" (by which he meant, according to historian Carl Becker, self-evident to some but not others). In truth Jefferson had no authorities to cite for his claims or any supporting proof based on evidence and logic. He spoke from intuition, common sense, and Christian tradition, from sources that eluded the physical senses.

Philosopher Jeremy Bentham soon pointed out the logical flaw at work. Rights-claims were mere nonsense, the empiricist Bentham blasted. Yet the popularity of rights and rights-rhetoric would only spread and gain momentum, Bentham (and conservatives such as Thomas Carlyle) notwithstanding. Proponents of the new rights and expanded liberty also had little trouble pushing aside John Stuart Mill when in the mid-nineteenth century he explained how individual rights were legitimate only when they derived from and helped promote the common good: Rights were the contingent products of social consensus, Mill stated, not timeless individual entitlements, and they needed to prove their social worth. Other writers disagreed (German idealists in particular), but by the age of Mill the argument meant little. Individual rights had acquired a secure cultural status.

The origins of rights and rights-talk are important to revisit in any effort to make sense, morally and intellectually, of our oldest task. Much of our moral thinking and talking is now based on individual rights. Further, they wrap around and help legitimate a worldview in which humans are supreme and in which they are best understood and treated as autonomous individuals, not as members of social and natural communities. When we fail to appreciate the social origins and contingency of individual rights we give them special moral status, and do so even as they fail to provide, standing alone, a sturdy moral frame for making sense of interdependencies and the common good.

The limits of liberty and equality. This second category of cultural elements connects, as noted, to the third category, having to do with our individualism and our particular enthusiasm for liberty and equality. Our long, slow recognition of human rights has brought many benefits, of that there is no doubt, with more work still awaiting. Before the era of human rights there was the Renaissance-era emphasis on man-as-the-measure, a humanistic awakening that similarly unleashed vast creativity. Yet, as in the case of our emphasis on objectivity—on sticking too much to facts and reason—our individualism can be and has been taken too far. Excessive individualism weakens senses of collective identity and collective fate, particularly individualism encased in broadly defined rights. It weakens the recognition that communities as such are legitimate sources of social values and expectations. It fuels calls for morally neutral states—a fictitious and misguided ideal—and for ever-looser limits on individual choice. Today's libertarian thought is a form of fundamentalist dogma suited only for the selfish and short-sighted, with politically far-left versions only marginally better than those on the far right.

As explored in chapter 4, liberty deserves special emphasis among competing public values only when it is defined in its fullness, only when the concept covers not just negative individual liberty (the individual's freedom from restraint) but also positive and collective forms of liberty, particularly the liberty of people in combination to take responsible charge of their shared fates. Even as to negative individual liberty, it has over time come to focus (as it did not in the past) on freedom from *state-imposed* limitations, even as so many constraints on individual life now stem from market-based private power (a reality recognized by Progressive Era reformers a century ago). As for equality, we struggle to think and talk clearly about it. We struggle to recognize that equality is not an independent moral principle but instead a fragment of one. As such it is a moral claim that can produce good only when embedded in a larger, sound moral vision, as it was in early centuries. (The same complaint can be leveled against the general idea of rights.) Formal equality means treating like cases alike—nothing more than this—and it cannot tell us when two cases are alike. It cannot on its own tell us when we should ignore differences among people and cases, thus viewing them as alike, and when instead differences are morally significant. Even when combined, individual liberty and equality as guiding beacons fall far short of the kind of moral order we need to come to terms with our ecological plight. In moral terms we resemble the arm-wrestler who spends years strengthening one limb while letting the others deteriorate.

When it comes to ecological decline our individualistic view of the world implicitly points a finger at people as individuals, tracing environmental ills to the free choices that they make. How else would the ills arise if individuals are, in fact, the prime components and movers of society? If that is true, if individual choices are at the bottom of all things good and bad, then the evident solution is to encourage people to make different personal choices, better ones. Today's real situation, though, is quite substantially different. Few environmental problems are entirely or even largely caused by voluntary choices that individuals make in their private lives. Further, hardly any significant problems can be well addressed by encouraging different individual choices without more. The problems are systemic, rooted in institutions, structures, and patterns of communal interaction.

Competition and the market's domain. Into a final category of troubling cultural components we can usefully place our implicit sense that progress arises largely if not entirely from competition and individual striving, particularly in the market. As historian Daniel Rodgers records, by the turn of this century "no word flew higher or assumed a greater aura of enchantment than 'market,'" which came to stand in the popular mind "for a way of thinking about society with a myriad of self-generated actions for its engine and optimization as its natural and spontaneous outcome."[3] This faith in self-seeking competition seemed to rest solidly on the economic theory of Adam Smith and other classical and neoclassical economists and on the natural-selection theories of Darwin and his successors. In fact, Darwin had little confidence that competition and natural selection led to progress in any meaningful sense. Those who survived in nature were simply better at survival; little more could be said. Smith, too, never claimed that market processes inexorably brought economic gain—they merely could do so, and often did. Like most of his age, he presumed that even profit-seeking actors in the market would be constrained by Christian ethics.

Biologists now know that evolutionary fitness for humans and many other species has come about as much through cooperation and deference to communities as through individual competition. Selfish individuals may beat out altruistic ones in head-to-head competition but the "iron rule" of genetic evolution, says biologist E. O. Wilson, is that "groups of altruists beat groups of selfish individuals."[4] It is similarly relevant that individual competition can in fact yield evolutionary changes that harm a species (particularly, it seems, physical competition among aggressive males).

Looking ahead, we surely need a heightened recognition of the ways competition can prove wasteful—economically, socially, and ecologically—as well as beneficial. We need a sense, too, of the ways it is morally corrosive and prone to undercut the social norms and structures upon which it depends. As conservative Richard Weaver has proposed, competition and liberty need tempering with the age-old counterbalance of fraternity, "the ancient feeling of brotherhood [that] carries obligations of which equality knows nothing."[5]

Rising above these four categories of cultural flaws—towering above them, some would say—is the implacable reality of nature's limits. Our planet, to be sure, receives daily infusions of energy from the sun. And we can admit, too—indeed celebrate—human cleverness in manipulating the earth's physical matter. But it is utter folly to charge ahead, bound by bound, without keeping an eye on the horizons, without pondering our plight and planet from the perspective of the stars. Our atrophied morality is in no way more evident than in our selfish conceits that cleverness and competition have, *mirabile dictu*, lifted the fetters and reserves of shared virtue.

———

If I am right in my claim that our ecological ills are rooted in problematic aspects of contemporary culture (perceptions, beliefs, and norms), not just in physical factors and population, then we have need of significant cultural reform:

Collectively we require a *better-grounded vision* of what it means to live well in nature to help us head in that direction. We do not now have such a normative vision, not in common currency, and for reasons already sketched. Such a vision would be grounded in morals and not merely facts and reason; it would embed us in natural systems in much the way other living creatures are embedded, challenging our boastful claims of independence; it would present us, in important ways, as parts of larger communal wholes, ecological as well as social, and not merely as autonomous individuals; and it would stress needs for us to cooperate for the common good, even as we retain and reward competition and creativity within morally prescribed bounds.

Somehow we need to *gain the wisdom* that we are not merely solo beings, adrift in a purely physical, morally empty world and competing for places to build homes and plant crops. Somehow we need to learn (relearn in part) that we cannot understand components of a landscape—or of an ecosystem or the world as a whole—without comprehending the whole as such, a line of organic reasoning readily traced from ancient times through Spinoza, Hegel, John Dewey, and others. Similarly we need to grasp how emergent properties so often arise when parts cooperate and evolve over time, giving rise not just to greater capacities but to capacities different in kind.

Above all, we need to find ways to *talk collectively* about moral values, ways that make it sensible and authoritative for communities as such to settle upon values that guide and constrict the lives and choices of individual community members. Values need to become something more and different than personal preferences. (Say it again.) Likely this means renewed talk about the good life, about good character, and about how both are linked to the honorable fulfillment of communal roles. At the moment (and as varied critics have said), we are woefully short of moral language that goes beyond negative liberty and individual equality to reach the higher grounds of cooperation and long-term community health. Good morals, good cultural controls, are not just out there waiting to be discovered, certainly not through reason and empirical data collection alone. Instead they are tools that we craft, as the value-creating beings that we are.

———

My call for cultural changes is not merely a call for reformers and activists to absorb these ideas and then use them as they develop environmental policy reforms and press for acceptance. It is more radical than that. Effective environmental reform needs to focus directly on cultural values and assumptions. It needs to find ways to foster reform at that foundational level. The failure to recognize this, and act upon it, explains better than any other factor why environmental reform efforts have run aground and why environmental activists are so often spurned. It just won't work to appeal to people today where they now are, to speak to them in ways that make immediate sense given the deficiencies of modern culture. To the contrary, true reform needs to lure people into becoming something better than they now are, encouraging them to want and work for modes of living, understanding, and valuing consistent with a flourishing, enduring civilization.

Environmental reformers need to recognize that the challenge they pose to society is, and must be, more profound and demanding than the reform steps aimed at ending racism and promoting marriage equality. Hard as those social challenges have been, reform efforts have worked within, and have not needed to question, main elements of the modern worldview, morality included. Reformers could accept (and have) the idea that moral value resides only in humans and that we are best understood as autonomous individuals. They could accept, and have, short-term time horizons while doing little to shake our towers of human arrogance. For the most part, civil rights causes have wanted to help excluded people fit fairly and fully into the modern system, not to challenge the system as such, certainly not the capitalist market and dominant power structures. The environmental cause, in contrast, requires vastly more in the way of cultural and institutional change. It threatens power structures and business as usual far more than that, or certainly needs to. Indeed, to succeed it must pose a frontal challenge to the dominance of the very moral standards that civil rights causes have raised high in battle.

This inquiry, by its end, brings together the various components of a new, more land-respecting and enduring culture. In doing so, it identifies the key normative considerations useful in distinguishing between the legitimate use of nature and abuse, though without (for reasons explained) knitting them into anything like a ready-to-use test. More broadly it presents a call for a new direction for environmental reform, one that seeks first and foremost to promote a more land-respecting culture. The final chapter includes an overall strategy, identifying what needs to change in today's culture and how reformers might best talk about that change.

————

A few words are needed to explain the book's recurring use of American examples when discussing global cultural problems. The United States was founded upon, and during the heyday of, Enlightenment thought. By the end of the eighteenth century (far more than in 1776) the new nation's identity was increasingly bound up in ideas of liberty, individual free competition, and the rising ideal of equality, principles that fueled distrust of government. Over the generations the nation came to exalt its mixed immigrant heritage and diversities of religious views, particularly among Protestant Christians. What held the nation together, given this mixed heritage, was not ethnic or religious bonds or any extended shared history. Instead the nation gained identify chiefly with political principles. It was a republic if not a democracy, a place of liberty and opportunity, a place where economic and social barriers were knocked down and individuals could rise as high as energy and abilities allowed. The nation's dominant institutions, private property and the market above all, reflected these cultural values. Progress came by physical expansion and growth, by continued protection of individual rights, and by the free play of individual competition.

This heady mix of freedoms and opportunities has brought great gain to the United States, particularly in economic terms. It has also brought heightened respect for outcast and subordinate elements of the population. But these same cultural values, we need to see, also play foundational roles in ongoing ecological decline. They are cultural values that, in their place, yield distinct benefits but that, taken too far—as they have been—bear increasingly ill fruit. Today they underlie political impasses that are more severe than the nation has ever experienced apart from the days before the Civil War. They largely explain why the United States has become such an assertive nation internationally, not just dropping more bombs and dispatching more missiles over the past sixty-five years than all other nations combined but refusing to sign numerous international agreements (for instance, the Convention on Biological Diversity), ones that nearly every other nation has seen fit to sign.

My cultural criticisms, then, apply with particular force in and to the United States, which likely faces a greater need than any other nation to adjust and reassess its founding cultural elements. To be sure—and it needs emphasis—the United States has made progress in dealing with significant environmental ills. It proudly thumped its chest when the Soviet Union collapsed and the world saw how much ecological degradation had occurred behind the Iron Curtain. The fall of the Soviet Union, however, far from being an impetus for the United States to do even better on ecological issues, seemingly became instead a justification for quashing the very collective forces that accounted for the nation's environmental progress. Libertarian and free-market enthusiasts were quick to assert that the better environmental record of the United States somehow had to do with its embrace of the free market: the United States was capitalist, the Soviet Union was communist, and capitalism took better care of the environment. Economic growth, higher under capitalism, somehow magically transformed into better environmental outcomes.

A more careful review shows that this narrative captures precious little truth. Rhetoric aside, the Soviet Union possessed a capitalist economy, every bit as devoted as Western nations to industrialization and to the reinvestment of profits into productive plant and equipment (the classic definition of capitalism) as Western, more market-based versions. The main difference was that Soviet-style capitalism was largely unchecked by popular insistence on environmental responsibility. That is, it lacked effective popular calls to control pollution and ecological degradation. One need only look at the often-dismal environmental records of US-based global corporations around the world to see how they would likely behave at home if binding environmental laws didn't hold them to higher standards. The true story is complex and knotty, but there's little doubt that the better environmental record of the United States had far more to do with citizen-led reform efforts and with the enactment (over vehement business resistance) of major environmental laws than it did with any mysterious market forces. To be sure, a wealthier nation can invest more in pollution control and other environmental measures. But the market itself does not motivate such investments; to the contrary, it encourages cost-cutting and profits.

Perhaps the biggest practical challenge we face, institutionally, on the path to environmental sanity is the work of embedding the capitalist market into a healthy, moral social order and into natural systems that remain fertile, productive, and biologically diverse. As we are learning—as we should certainly know by now—the market is a valuable servant but a terrible master. The good news is that public faith in the market is sagging, and for many sound reasons. Its star is tarnished, much as the star of science lost luster in World War I. Many feel it, even if subconsciously: the time has come for a new way of organizing ourselves, a new way of focusing and expressing our hopes.

To escape the tyranny of the present, to stand well back of modern culture and to view it critically, is to see major flaws in our worldview and to see too how these flaws largely account for various contemporary ills. Fortunately, it is also to see ample opportunities for beneficial change, change that can best begin, as perhaps progress has always begun, with a clear-sighted, critical look at the present.

ONE

Composing the World

To get a clear sense of our place in nature, a promising place to begin is with the natural world itself and with an effort to see, hear, and otherwise sense what the world contains. Of what is reality composed, in terms of its visible and invisible components? We commonly let scientists to tell us about the world, supplementing what they say with everyday experiences. But science's tools have limits to them and, even using science, our understanding of the natural world, more than we realize, is considerably influenced by modern culture. In important ways it is something we compose, certainly our understanding of it is. A clear-sighted look at our culture—at our present-day predicament—needs to get at this bottom or foundational level of our existence in the world. It needs to expose how we commonly think about the world's constituent elements and about our capacity to comprehend them. For good and ill our behavior depends on it.

This chapter is about intellectual foundations and involves (by way of warning) a lot of digging and below-ground work. A construction metaphor seems apt here because a new cultural order cannot simply be added atop of what we now have in place. We can't get to a new awareness of our existence, we can't make basic shifts in our moral thought, without reengineering that begins at the lowest levels, cutting away in one place, adding new support elsewhere. Central elements of today's culture need to take on new forms. Proposals to revise them— proposals, for instance, for new moral schemes and new ways of understanding—will (and do) meet resistance, and for the very good-sounding reason that they are misguided, because they clash with reality as widely perceived, because they fit badly on the cultural foundations and building blocks we now have in place. Reform in the public arena, in short, needs to draw attention to these foundations and to the work of rebuilding them.

Making Sense of the World

The task of identifying and cataloguing the world's many parts is hardly an easy one. The natural order is immensely complex. Our connections to it are mediated by our senses and hence our bodily functioning. This first step in the inquiry, in the quest to see the world more clearly, thus needs to attend to the ways we gain knowledge about the world, gauging in particular the constraints on our knowledge that stem from our limited sensory capacities. It needs also to illuminate how our brains, overloaded with potential stimuli, necessarily screen data and pay attention only to what seems most important. As for what we are able to see, hear, and feel, our brains must assemble the countless data points into images of the world. This subconscious image-making process poses another filter on our ability to gain direct, reliable knowledge about the world. In significant ways our conscious interactions with the world take place with these self-constructed images, not just with the physical world itself. We especially rely on our brains to detect or attribute causal connections among events. Our brains play critical roles, too, in making sense of larger wholes and discerning purposes or motivations. Ultimately it is our inner selves—our minds, hearts, or souls—that pass judgment on the goodness or badness of all that we perceive and do.

Anyone who has dipped even a single toe into the ocean of writing on philosophy will know that these observations on our creaturely plight have, since ancient Greece and before, provided the stuff for philosophic contemplation. And the flow of speculative writing on them (by philosophers and others) shows little sign of easing up, even as particular lines of thought lose favor for a time only to rise again in revised form.

- Of what is the world composed and does it operate according to a purpose or plan?
- Are we able as finite humans to perceive reality directly or do we instead see only indirect images or imitations of it?

- As for what we see, are objects wholes in and of themselves or are they merely pieces and parts of larger entities?
- Why do objects take the shapes and forms that they do and display certain traits and characteristics and not others?
- What provides the motivational forces behind physical motion, birth, and death?
- As for our senses, how reliable are they, and does the world contain entities or spirits that escape even our closest attention? How can we distinguish between dreams and reality? Indeed, can we even be sure that our sense impressions aren't triggered by spirits playing tricks on us?

Philosophers over time have asked these questions and others like them and come up with a dizzying array of answers, some that sound sensible today, others that do not. Particularly influential for many centuries were the speculations of Aristotle, who concluded that all objects in the world could be explained in terms of various causes. In his view, each object had a final cause (one of several), which prescribed its purpose or its final state of existence. He believed also that all living things contained animating spirits or souls of one or more types. Humans rose above other life forms because they possessed all three types of souls—the vegetative soul responsible for metabolism and growth; the animal soul responsible for senses, motion, passions, and instincts; and the human, rational soul that empowered people to think, reason, and speak. Aristotle's causes and souls seemed to reside among the world's component parts, even though they lacked materiality and were present and knowable only when embodied in particular tangible things. In taking this stance he clashed with Plato before him, who famously asserted that the physical objects we witnessed were merely imprecise images—flickering shadows on the cave walls, as he famously put it in his *Republic*. In some unknown way, Plato asserted, these images were derived from intangible but nonetheless real Forms or Ideals, which were the true stuff of reality.

Until recent centuries, philosophers typically assumed that the world around us arose from or was put in place by God or, at the least, that God or a similar intelligence kept reality together and ordered its parts. Also running through the philosophy's history, beginning with Plato if not before, has been a distrust of our sense impressions, a belief that sensory perceptions are mere opinions and unreliable. Given this unreliability we could not really know the truth via our senses; truth was relative, as early Sophists had urged, a matter of probabilities based on individual judgment. This skepticism gained intellectual popularity in the ancient world even as other philosophers—Socrates and Plato among them—relentlessly pursued knowledge that transcended mere opinion. For many, the world gained order and meaning by comparison to the ideal of the Cosmos, a distinctly Greek vision combining order, structural perfection, and beauty against which the experienced world could be compared.

For our purposes we can take up this tale of speculation about nature and human capabilities with the Enlightenment thinkers of the eighteenth century. It was their varied contributions that soon exerted considerable, lasting influence on educated minds of the West.

Enlightenment thought built upon the Renaissance-era emphasis on the human as a capable, individual creature: The human was a worthy being, not merely a pathetic, fallen figure whose ultimate plight depended solely on God's mercy. It was a new, more positive view of humankind, conspicuous in works by Erasmus, Machiavelli, and others, a view that heightened confidence in human powers. People could learn directly from nature without depending on divine revelation and with their unique powers of reason could think critically about it.

Writing in the early seventeenth century Dutch philosopher and mathematician René Descartes brought new attention to critical reason and its intellectual benefits with his prominent uses of systematic doubt and with his insistence that all factual claims be based on hard proof. His contemporary, Francis Bacon, brought similar attention to the potentials of empirical data collection and inferential reasoning as tools for gaining direct knowledge of the world. Astronomers played key roles in this as they collected data and reasoned their way inductively to surprising conclusions about planetary motion and the layout of the universe. As is commonly known, Galileo would run afoul of the Catholic Church for his scientific work. His chief sin, though, was not that he put the sun at the center of things (bad enough) but that he denied, *contra* Aristotle and church teachings, that objects in the world were shaped and guided by inherent causes. There were no such causes, Galileo countered, and perhaps no intangible elements or spirits of any type. All was flux, all was a matter of physical stuff interacting according to planetary physical laws, laws that Isaac Newton and others were reducing, step by step, to mathematical form.

The Enlightenment era that rose with Galileo and others built on this greater faith in human capabilities, in human reason in particular. It built on the new belief that knowledge came most reliably from human inquiry rather than tradition or revealed religion. Sound knowledge came from empirical data collection using human senses and logical inferences from the resulting data. The discovery of predictable patterns in physical nature—nature's laws, they were termed—provided strong evidence of these human powers. It also lent support to a longstanding minority view that nature was nothing more or other than physical stuff, moving about according to laws that were exceedingly complex but ultimately knowable by humans. Nature was one massively complicated clock with interrelated parts, highly complex but ultimately not mysterious.

Impressed by these intricate interconnections, Dutch philosopher Baruch Spinoza contended that the world as a whole was the only fully independent substance in existence. It was the only primary substance within the Aristotelian frame of understanding. As such, only the world as a whole could be understood in isolation; everything else could be understood only insofar as it participated in the larger whole. As for what held this whole together and gave it order, Descartes took the traditional stance, as did Leibniz, John Locke, and others of his century: God was at work, they believed, linking all into an organic whole. Yet, even as they gathered data and generated their mathematical formulas, philosophers retained doubts about the fullness of what they could perceive and thus know. Human perceptions were inevitably limited, as John Locke asserted. This likely meant that much lay beneath the surface of things, outside the reach of human senses. The sensing human was a powerful creature but far from all-knowing.

By the mid-eighteenth century, the golden age of the Enlightenment, faith in human reason had risen high, particularly among the educated. So had the belief that empirical data collection and analysis, whatever their limits, could give rise to knowledge that was more reliable than knowledge gained from any other source. Out of this mix arose a heightened faith in science and scientific inquiry along with a belief in reason-based education. Out of it, too, came the sense that science's possibilities could be unleashed only if scientists were allowed to operate free of external control. It was therefore essential for science to shake off the constraints on independent inquiry imposed by aristocratic orders and by the church. Scientists had to gain their freedom.

In time the Enlightenment project came to emphasize political freedom as a necessary condition for progressive change. For more radical figures, that freedom needed to take the form of individual liberty made available on an equal basis. A few radicals were willing to go even further and to question the entire idea of a divine moral order in the world. Perhaps this too, they proposed, was simply another religious myth handed down for centuries that reason should wipe away. Nearly all leading figures, however, were inclined to retain a lofty moral post for humans, far above other creatures. Humans were a distinct form of life on earth, at their best guided by reason and possessed of special powers. They were uniquely capable of making sense of the world and challenging superstition, traditions, and aristocratic conceits. They were also capable of bringing about progressive change by means of objective inquiry and action. With facts in hand and careful reasoning, leading intellects could cast light onto the road of progress and direct others down it.

Implicitly in some cases, explicitly in others, the Enlightenment project was thus guided by a utopian vision of a perfect or near-perfect world. It was a utopia that would arise through human effort—through planned effort, most believed, but perhaps (as others would speculate) even through unplanned efforts, through the seemingly selfish actions of liberated individuals as they went about creating, buying, and selling in an unfettered marketplace.

Reason and Its Limits

The Enlightenment era's successful elevation of the fact-collecting, rational human to a superior status in the universe helped unleash revolutionary movements. It prompted calls for liberty and equality and for new forms of economic and industrial creativity. Clearly a new confidence was in the air. Progress seemed possible if indeed not inevitable, at least when old forms and structures were cast aside. Knowledge came through objective inquiry—for many, its only reliable source. As for the calls for liberty, they quickly challenged the powers of both state and church when it came to imposing legal and moral limits on what individuals could do. Particularly in the United States and among European nations bordering the Atlantic, the call grew loud to recognize and honor individual rights. Confidence was high, humanity was the principal measure, and individual initiative was displacing institutions to become the prime force for progressive change.

And yet, even as Enlightenment ideas and sentiments gained such popular force, the most important thinkers—including those most responsible for the new wisdom—were suffering doubts about them. Some simply couldn't get beyond the fact that human senses were not completely reliable. How could we exalt facts and reason from them to sound conclusions when facts about the world around us came from our senses and our senses gave us distorted images? Just as troubling was the recognition that fact collection and reasoning weren't yielding support for the existence of a moral order in the universe, divinely imposed or otherwise. Ideas of goodness and virtue weren't things that could be touched, heard, or smelled. There was no physical evidence of their existence that science could collect and study. Reason, too, could achieve little on its own, in isolation. Reason was a tool to organize facts and ideas and tease conclusions from them. In the case of reasoning about morals, it needed to work with first principles or axioms about goodness or moral value. Without them it couldn't get started.

Some philosophers of the era, convinced about the unreliability of sense perceptions, would largely wipe their hands of everyday concerns and decide that we simply couldn't know anything with certainty. But this was an impractical stance, unhelpful in daily life and unlikely to gain wide support. Individual initiative had been let loose and industrialization was beginning. Achievement and wealth came to those who charged ahead, not those who doubted human powers.

A more constructive response came from Enlightenment philosophes—David Hume perhaps best remembered—who could see that moral judgments derived, initially at least, from feelings or sentiments more than they did from rational thought. We judged right and wrong based on them or based on intuitions, Hume contended. Reason entered the picture to screen, modify, and elaborate these sentiments. Contemporary Jean-Jacques Rousseau largely agreed. Guided by their inner natures humans were innately good, Rousseau provocatively asserted. Moral behavior came to them naturally so long as society didn't get in the way.

Sentiment-based approaches to morality made sense to many public thinkers. Others were more prone simply to seize on particular principles and proclaim their truth rhetorically, as Thomas Jefferson did with his individual rights. Writing at the turn of the nineteenth century, Jeremy Bentham gained followers by contending that public policies should be based chiefly on the promotion of human happiness—a moral principle similarly embraced as an unprovable axiom. Policies should promote overall human welfare, he contended, not be tailored to respect human rights. Bentham's simplistic reasoning initially encountered headwind because he seemed to sanction selfish behavior. He seemed to challenge the common virtues of honesty, purity of heart, and Christian charity. Sentiment-based approaches seemed no more intellectually satisfying. Surely morality was not just a matter of individual feeling, as Hume seemed to imply; surely it was made of sturdier stuff of some sort, more binding on people.

So long as people largely embraced a traditional moral order this inability to find a solid grounding for morality could be overlooked. But by the turn of the nineteenth century, and increasingly thereafter, reports came in from around the world about societies and cultures with widely varied ideas of rights and wrong. Indeed, hardly any action that Western Christian society viewed as manifestly immoral was not, somewhere in the world, embraced by people as a positive good. Given this variety, how could morality be based simply on intuitions or sentiments without risk of degenerating into mere group, or even personal, preference? And how could it be based on making people happy when happiness came in such widely disparate ways?

Despite these anxieties, the Enlightenment push for reason-based progress continued apace. Particularly expressive of it were the varied efforts in the early nineteenth century to design and construct utopian societies and the related calls for institutional change. Some reform visions took on socialist leanings. Others were shaped by religious enthusiasms, nominally based on scripture and divine inspiration but routinely calling on capable people to do the creating. In their writings, Karl Marx and Friedrich Engels would express a similar Enlightenment-based belief that progress could, and likely would, come about and that human efforts—class-based more than individual—would make it all happen.

In retrospect, the strongest line of reasoning forged in the liberating furnaces of Enlightenment fires was the belief that individual humans, vested with liberty, could bring about progress through their economic enterprises, without much need for overall design or indeed for any moral vision other than freedom and equality. Adam Smith in his *Wealth of Nations* set the tone, even as he personally remained part of an older generation, even as he assumed that the traditional moral order would remain in place and contain the selfish acts of market participants. Soon, free-market apologists such as Thomas Malthus would use their own economic reasoning to defend the market and to resist social-reform efforts that sought to change its economic outcomes, particularly its mass poverty and economic inequality. Malthus issued his infamous writings on population not as a call to contain population growth, but instead to explain why the poor would always be with us and why Christian charity, to say nothing of regulations of business, were misguided. Paradoxically, then, the Enlightenment gave rise to offshoots that soon ended up in direct conflict—on one side, efforts using facts and reason to promote citizen- and government-sponsored progressive reforms and, on the other side, efforts to defend the market against such reforms so that the market itself could bring progress. Enlightenment objectivity, it was turning out, led in differing directions.

The Acids of Modernity

The nineteenth century was a time of exceptional economic growth. For most of the West it was also—after 1815 and excluding the American Civil War—a time of exceptional peace. Signs of progress were everywhere, even as the economy created losers as well as winners and even as increasingly powerful forces seemed to push people about. As for the nagging intellectual flaws of the new worldview, they continued to fester. In the Enlightenment creed humans were the measure, the world's uniquely rational creatures, and science followed a path that would one day rid the world of mysteries. But decade by decade it became harder to retain this faith. Science and reason gave little cause to think that the world was created for humans, even if they did possess powers unique among life forms. Many human traits were also present in other living creatures, as rural dwellers had long known. By the early nineteenth century biologists were speculating widely that humans had somehow evolved from the apes, a claim Rousseau had popularized earlier. The linkage was freely discussed, even though scientists could see no mechanism to account for it.

Meanwhile, archaeologists and geologists were turning up evidence that the world was far older than the biblical account. From remote lands anthropologists and world travelers were dispatching stories of alarming cultural practices, present as well as past, which seemed to cast doubt on the alleged superiority of humankind. In some reports, originating from the inner city as well as from afar, humans seemed hardly less vicious or irrational than other primates. As for the evolving social order, the ill effects of industrialization and the ruthlessness of financiers raised similar questions about the overall trajectory of humankind. Economic progress was plain enough but it seemed to come at the expense of public morality. Some new, much stronger moral glue was needed if destructive forces were to remain in check.

Looking back, it is common to point to Darwin, Freud, and Einstein as the most prominent figures in ushering in the new, disorienting age of modernity. The twin pillars of the former age were the Enlightenment faith in science, reason, education, freedom, and progress—all proceeding hand in hand—and the still-dominant Christianity-based moral order, which supplied the needed limits on individual striving. By the early twentieth century both were under siege.

In their writings on natural selection Charles Darwin and Alfred Wallace came up with the missing mechanism to explain how evolution might happen, greatly increasing the theory's credibility. Humans were not just closely related to other living creatures, it now seemed. They arose in the same natural manner and, given their lowly origins, had no nature-based claim to special moral status. Evolution, moreover, had been going on for millions of years. This apparent fact, added to evidence from geology about the earth's ancient age, meant that the biblical account of creation was not literally true. Earlier centuries had not much concerned themselves with literal truth; biblical narratives were valued more for moral instruction than historical insight. But for the post-Enlightenment mind literal truth had become a litmus test. The work of Darwin and the geologists appeared devastating.

Just as Darwin's work gained potency when combined with geology, so Freud's writings on human irrationality gained strength from reports from anthropologists about deviant human practices around the world. If one could believe these accounts of extraordinary cultural practices, humans were not the rational beings that the eighteenth century had claimed. Reason, it increasingly appeared, was mostly a tool of the passions as Romantic critics had contended. In Freud's account, much human action was guided by subconscious forces over which the individual had little control. Even as to conscious thought and choice, different peoples around the globe seemed to draw widely different conclusions about morality and how people should behave. Given this evidence, how could one retain the Enlightenment-era faith?

One explanatory possibility was that certain peoples were simply more primitive than others. In due course, these primitive cultures would by steps come to resemble the mannered, sophisticated order of the developed West: Cultural evolution just needed more time. It was an appealing line of reasoning and quickly seized upon. Evolution, it seemed, took place in human societies and in institutional structures as well as in nature. Just as advanced humans rose above the apes, so too primitive economic and social systems around the world would increasingly mimic the republican, rights-based systems of the West.

Language about primitive-versus-advanced and theories about inevitable economic and social evolution would continue to draw adherents for generations, even as evidence from the field provided ambiguous support. But the popular embrace of this hopeful view was never full enough to overcome mounting doubts and anxieties, particularly as the older moral order seemed to disintegrate further. How could one believe that primitive societies would evolve toward greater sophistication when the most sophisticated societies were becoming more vulgar and faithless? Belief in progressive evolution took an especially hard hit with the coming of World War I and its unprecedented forms and rates of slaughter. Only rational, technically advanced societies could engage in killing on this scale. Given such bloodletting, how could meta-narratives of inevitable progress remain plausible?

Then there was Einstein and his claims, leaking out from Switzerland, that even the physical world around us was not quite what we thought it was. Few could understand the details of his theories, but they didn't need to understand them to sense the tremors. Even before Einstein, scientists spoke of realms of tiny matter that were far beyond human perception. Items that appeared solid were, apparently, anything but. At the other end, astronomers were talking about a universe beyond our sensory perceptions if not beyond our comprehension. Einstein thus only added to the mounting evidence when he claimed that the laws of physical motion as we perceived them did not always pertain. Our senses didn't fully track reality.

For those who pondered the matter, this new scientific evidence added a renewed element of mystery to nature. Some gain could come from that revival, but the dominant effect was to push humans down another notch, to sap a bit more of their sense of being special, to sap their belief that ordinary sensory perceptions and reason supplied an accurate account of the world and reliable tools to interact with it. Softening the blow to the collective human ego was an accompanying sense of awe at the prowess of the scientists themselves. Ordinary individuals might have limited capacities, but humankind collectively with its top scientists still wielded sharp intellectual swords.

All of these forces came together in the 1920s to give rise to a powerful sense of unease among the educated. Religious belief was hard to retain except by deliberate personal choice—a distinctly inferior grounding for faith. The central elements of the Enlightenment view of the world also seemed shaken, even if people in their quotidian affairs gave little sign of it. Most troubling for those paying attention was a key problem identified a century and a half earlier, the lack of any way to ground morality once religion was pushed aside. Were the old moral verities simply myths, as reason and anthropologists seemed to show? Along with this instability in moral foundations came a loss of any real sense of purpose for human life. Did meaning still linger out there somewhere, was one of evolution's products charged with a special role in the world, or was even human life nothing more than an aimless journey of birth, struggle, and death?

Few writers expressed this existential plight better than Joseph Wood Krutch, sometime journalist and drama critic, later an especially thoughtful writer on landscapes and the American Southwest. In a bestselling work in the 1920s, *The Modern Temper*, Krutch offered a sharply pessimistic status report on Western culture in its most advanced forms. Morality, Krutch lamented, was in fact dead as anything binding on society as a whole. For the sensitive individual modern science had created a spiritual crisis, Krutch claimed. A sensitive spirit could live and thrive only in a world where moral values had real existence, a world in which right and wrong were objective forces, love was more than a biological impulse, humans possessed real powers of voluntary choice, and reason was more than a pawn of emotions. Science, it seemed, had disproved all of these points, or rather shown them to be illusions, leaving the sensitive individual adrift in an alien if not meaningless world.

Krutch's pessimism would be outdone during the depths of World War II by an extraordinary critique assembled by German writers who had fled the coming horrors. In *Dialectic of Enlightenment* Theodor Adorno and Max Horkheimer blamed the mass destruction of the 1930s and '40s on the Enlightenment-era tendency to detach the mind from the body, its tendency to exalt reason and objectivity and, in effect if not design, to legitimate the pursuit of self-directed goals. Particularly in the case of the most successful competitors, these philosophers claimed, the self became more and more driven by the individual ego, which sought to pursue individual pleasure with little concern for the effects on others. The self not only differentiated itself from others—dissolving all senses of fraternity and organic unity—but pushed aside culture, religion, morality, and community oversight in the quest to get ahead. Emptied of values and responsibility, drained of moral sentiments, the ego reached the stage where it sensed it could do anything and believe in anything. Only such an ego, with reason simply a tool of desire, could see mass killing as a legitimate tool of domination.

Krutch proposed that people respond to the emptiness of modernity with a grin-and-bear-it, stoical calm. It seemed the most dignified alternative for those who could see how science and the fruits of Enlightenment philosophy had drained the world of morality and ultimate purpose. A more common response, in dreams if not in life, was the one portrayed in the writings of F. Scott Fitzgerald. The egotistical pursuit of wealth and pleasure, vividly exposed in *The Great Gatsby* and other tales, was not without its satisfactions and rewards. But, as Jay Gatsby came to experience, it was often an empty pursuit. Lacking a social and moral grounding, lacking a sense of purpose beyond self-gratification, lacking any sense of participation in a larger organic order, even winners in the competition for worldly riches had trouble keeping a life together, much less helping to sustain those around them.

In its most acidic form, modernism of the 1920s entailed a rage against order, against the entire idea of coherence in society. It involved a lawless attitude toward moral norms and cultural judgments, replacing them with an emphasis on the self and a ceaseless search for experience.

Ignorance and Sentiment

For humans to survive at their oldest task, for them to live in ways that sustain natural as well as social communities, the path that led to Krutch's writings—to say nothing of *Dialectic of Enlightenment*—is a path that needs substantial correction. For people to live well on land they must instinctively if not consciously understand their roles as members of what Aldo Leopold termed the land community. They must see themselves as community members with responsibilities as such. For that to happen, for people to act responsibly in and toward nature, they must strive to understand the natural world. And they must be willing to act on their best understandings, even as they remain cognizant of the limits on what they know and can know and thus of the virtue of acting humbly. In the end, stances must be taken, moral choices must be made, and decisions must arise based on best-available evidence.

In this light we return to the task of the chapter, to make sense of the world around us—to compose the world, really, in that we must inevitably make decisions about what it does and does not contain and then be prepared, as best we can, to engage the world on these terms. Among the lessons of philosophers and, in recent centuries, psychologists and other scientists is that we face limits as finite beings in learning about the world around us, in gathering and assembling data perceptions and forming them into understandings. Our abilities to sense the world are substantial but much less so than those of many other creatures, ones that see, hear, and smell far better than we do. Scientific tools, of course, greatly expand the sense perceptions of those who deploy them. But many tools are usable only by specialists and in narrow settings. The tools themselves have limits, and the amount of factual data awaiting collection is literally endless. Of the estimated 8–9 million species of life on the planet humans over the centuries have identified and named (but in most cases not researched) some 2 million of them. We could call our failure to know more simply a matter of too little time and effort, rather than a limit of some sort on our ability to know. Either way we remain ignorant, even the best of us and even when our knowledge is pooled.

We need to admit, then, that there is much we do not know about the universe. Perhaps in time we will learn it all, but we have no reason to take comfort from such speculation. Whether we will or will not learn all in due course is of no practical consequence today. Our plight is to face limits on what we know, to make do with what we have in knowledge and abilities. Psychologists have amply confirmed Immanuel Kant's assertion in the eighteenth century that what we perceive is in part the creation of our brains, of the screening we do and the image-building we undertake. The proper response, though, is hardly to throw up our hands and bemoan our incompetence. It is to be as vigilant as we can in observations, attentive to the ways we might be excluding evidence, and following the best techniques for generating durable conclusions.

In a sense, then, the limits on our sensory perceptions are among the easier to accommodate: they call for humility, diligence, and attentiveness. A greater challenge would seem to come from the possible ways of knowing that do not rely directly on our senses, ways of knowing based on intuition or feeling, based on dreams or spiritual insights. Are we wise to push them aside, to insist that factual knowledge can come only through our senses and reason? Can we instead, as people presumably have always done, draw upon these alternative ways of knowing, somehow making room for them?

These questions are rightly put at the beginning of the inquiry for they speak to our very ability to learn anything about the world. An understanding of our ways of learning—epistemology—is logically prior to a survey of what the world contains. Yet it may help to postpone drawing firm conclusions on these epistemic issues until we have paid more attention to the possible components of the world and, in particular, have paused to consider morality and its possible groundings. To limit our tools for learning to our five senses (as augmented by scientific tools) is largely to rule out any ability to sense intangibles in the world. It is to dampen the possibility of detecting not just spirits and deities, but senses of goodness and virtue, of purpose, and of transcendent morality.

For the present, we might sensibly avoid a firm stance on this core question of epistemology: Can we gain knowledge of the world around us through means other than the collection of sensory data? For evident reasons it seems dangerous to rely on intuitions and feelings, more dangerous still to act directly on dreams or spiritual insights. Yet there are many ways and settings in which it does seem sensible to make room for insights that we cannot fully justify by pointing to particular, sense-collected data. Many philosophers have contended as much, particularly those like Plato who have suspected that our sense impressions are unreliable. Given their unreliability, Plato contended, we were wise to make use also of our imaginative faculties, both poetic and religious.

For starters, the human body works in strange ways and our engagement with the world is mediated in important part by genetic codes that have evolved over millennia. Our genes and the related biological processes that guide behavior do not strip us of free will; we are hardly mere automata, guided by programmed instructions that merely respond mechanically to external stimuli. But we are nonetheless inclined biologically toward particular behavior patterns rather than others. These in-bred inclinations—sensible ones we might presume— can enter our consciousness simply as feelings or intuitions. Moreover, our memories are by no means perfect. Over time we can lodge lessons into memory somewhere, lessons that, when they return, rise up not as conscious recollections of past events, linked to precisely remembered, sense-derived facts, but instead merely as feelings or senses about a situation, or about what we should and should not do. Thought and feeling form a blurred continuum.

We might illustrate this difficulty of distinguishing between facts gained by empirical data collection and facts that arise otherwise— more intuitively or through sentiment—by considering the cases of the experienced horse trainer and farmer. Both have need to size up the health and vigor of what they see—of a particular horse or of a pasture or grain crop. The experienced professional in these trades draws upon available facts, gathered by the eye and ear, by listening, and, in the case of soil, even by taste. Yet her judgment on health or vigor can go beyond these identifiable facts; it can arise from a sense of the case, from an inner intuition that is hard to attribute to facts individually or even in combination. Put more generally, our processes of observing and thinking are not detached from feelings and from prior experiences, including prior feelings. Particularly in such cases it is not possible to draw clear lines between fact collection, assessment, memory, and sentiment. All are brought to bear.

By way of contrast with this inclusive way of gaining knowledge we can consider the philosophic school of logical positivism, an intellectual effort arising around the turn of the twentieth century that sought (better than preceding efforts) to reduce claims of knowledge to those that were provable objectively, excising all other claims. Logical positivists limited their claims of knowledge to empirically gathered data and to conclusions that could be proved through rigorous logic. All we can know, they asserted, was what can be proved scientifically, either through rigorous induction from physical data or by deductive logic. All else could not be known and was thus meaningless, including all normative claims of value and attributions of purpose or meaning. Within a few decades, even this core of proven knowledge would be further reduced. Later analytic philosophers pointed out that deductive reasoning did not really lead to new knowledge: it merely showed logical relationships among what was already known. And as for knowledge from empirical data collection, it was forcefully urged that induction from facts could never really prove anything for certain, given that new data could be found at any time that qualified or contradicted conclusions from old data. All that could be known was a matter of probabilities.

The rigor of this logical-positivist approach has value by highlighting the limits on what we can know with complete certainty. Yet as a way of engaging the world it is severely flawed, and its flaws are ones that, to lesser degrees, also afflict other demands that we limit claims of knowledge to what we can be prove objectively. Most plainly, the approach has no good way to deal with gaps in our knowledge, no way to incorporate ignorance into the calculus that precedes action. By design it excludes all sentiments and intuitions. Further and as adherents admit, it is a way of knowing that does not and cannot lead to any normative assessments; it contains no means to distinguish right from wrong, wise from foolish, or important from unimportant. In our dealings with nature, the data that await collection are essentially infinite. The logical-positive approach can easily lead to inaction or indecision while the endless data collection goes onward.

We have good reasons, in sum, not to cast aside sentiments and intuitions, even as we might rightly be suspicious of them and want to subject them to critical thought. Sentiment, as we will see, is an essential root of normative judgments of all types, including moral judgments. It is highly valuable also in dealing with gaps in reasoning. It can support a stance of caution. Similarly it can justify a willingness to assume that nature's evolved ways of inhabiting a landscape contain wisdom that we might wisely draw upon, even when we cannot empirically learn what it is.

Searches for Truth

Questions of epistemology are closely tied to longstanding questions about truth and what it means for a claim to be true. We'll have trouble constructing an apt understanding of the natural world—or anything else, really—without some sense of what truth means, that is, of when we can rightly conclude that a proposed fact is correct. Disputes about factual claims—whether claims are or are not true—appear regularly in squabbles over alleged environmental ills and policy options. Greater clarity of thought ought to help.

The ordinary sense of truth, when it comes to a claimed fact about nature, employs what is typically termed a *correspondence* theory: a claim is true if it corresponds to facts about physical objects, events, or relationships in the surrounding world. So familiar is this definition, indeed so obvious can it seem, that one might wonder what other definition could be used. In fact, philosophers have pointed to problems with the correspondence theory of truth, not as an ideal or goal when learning about nature but in terms of its usability, with whether it can regularly, or indeed ever, be achieved. Many reservations to it are based on points just covered, having to do with the limits on the reliability of our senses and what we can learn by using them. They have to do also with the active engagement of our brains in screening data subconsciously, and with our need, realistically, to bring massive amounts of data together to form simple images, a process that also involves distortion. Then there is the simple reality that the facts awaiting collection are overwhelmingly numerous and we must stumble on with woefully incomplete data. Correspondence is ideal, but what do we do when it eludes us?

For these and related reasons, philosophers have supported other definitions of truth, ones that are more usable and more consistent with the ways we actually engage with the world. The *coherence* theory of truth posits essentially that the truth of an assertion is a matter of whether it fits together with—coheres with—other facts that are already accepted as true. Is a new fact consistent and well aligned with existing facts so that they form a coherent pattern? More colloquially, does a new fact "make sense" in terms of what we already know?

A third theory of truth—the *pragmatic* theory—seeks to overcome the inevitable idiosyncrasies of individual observers and individual brains by paying attention to the consensus judgments of many observers. It also pays attention to the consequences of accepting a fact as truth and then acting upon it: Does the acceptance of the fact lead to consequences that are good or instead to folly and nonsense? In practice, truth is often best understood as the shared judgment of a people, who through their varied experiences come together to agree on the truth of something. Truth also is, in practice, what seems to work when put to use; an assertion is embraced when it leads to better results than alternative versions. Pragmatists do not give up on the correspondence theory of truth. It remains the gold standard. But they are willing to accept claims as true based on lesser evidence, based on social acceptance by relevant group members and on evidence that true facts are more likely than false ones to bear good fruit.

The pragmatic approach to truth holds great appeal for people who act in the real world. It is particularly appealing when it comes to dealings with nature given nature's complexity and the often-unlimited facts that could be gathered. Rarely can we describe even simple phenomena in nature in ways that do not, in some minor particular, fail to correspond fully with the phenomena (or, even more, that are incomplete). And there are often high costs involved in postponing action and in the data collection and analysis. Far more sensible quite often is to embrace apparent facts as true, and act upon them, and then to adjust our actions as new knowledge. From this practical perspective the pragmatic definition of truth can hold real appeal, especially given its claim that truth is evidenced by good consequences. This is a usable approach to truth, however, only when the allegedly true facts can be put to actual use and when the consequences of using them can be observed within reasonable time frames. Moreover, this definition of truth requires having in hand a sturdy normative standard for judging whether the resulting consequences are good or bad. Without such a standard of evaluation, consequences can be identified and described but cannot be assessed. A standard of evaluation isn't embedded in pragmatism itself; it must come from somewhere else. And the truth of a pragmatic claim depends upon the soundness of the chosen normative standard.

These three definitions of truth, though varied in key ways, are not so different that they cannot be used in tandem. Of course it would be splendid to have sufficient evidence and unbiased data-collection methods to generate truth that satisfies the demanding correspondence test. But particularly when dealing with nature out-of-doors, outside the laboratory, the correspondence definition erects too high of a standard. More sensible is to judge facts by a combination of the other two definitions. Does an alleged fact fit together well with what we already know or think we know? Does the fact line up well with the sense perceptions of many people who have sought to confirm or refute it? Does the embrace of the fact, in practical affairs, lead to good consequences? As a variation we might ask, using these lines of reasoning, not whether a particular alleged fact is true (in the correspondence sense) but whether, on balance, it makes sense to accept it as true in that way and to take action, even while continuing to collect data and remaining open to revision.

This approach on the issue of truth—making do with what we have and working by consensus—is particularly necessary when, as is often the case, inaction itself is a course of action; when failing to act on an alleged problem is itself a chosen response. To bring this practical reality into the assessment is to interject the issue of burden of proof: who has the burden of proving an alleged fact, and what is the burden? Burdens of proof are of considerable importance in struggles at the oldest task. The selection of a sensible burden raises moral and other normative issues. The issue is taken up in chapter 3 and returns several times thereafter.

The Whole and the Parts

These comments on our tools for gaining knowledge, and on the meanings of truth, provide useful background for the work of composing the world, for making decisions about what it contains and about the nature of being or existence. Our dealings with nature are much influenced by how we see the world and what we think about it. On this subject, the modern Western worldview differs vastly from many earlier and alternative worldviews, particularly views in which the natural blended with the supernatural in a world populated by spirits and unseen forces and shaped by myth.

In the modern Western view the world is chiefly composed of physical stuff, of atoms and their inner elements bouncing about in space according to patterns and rules variously predictable and chaotic. It is an old view, atomism, dating back to ancient Greece. The big questions that need assessment today are of two basic types. First, there are the questions of how all these tiny parts fit together and how we might best think and talk about their interconnections: To what extent are they better described as independent things or instead as parts of larger wholes? To what extent should we talk about the pieces as such, separately or as aggregates, and to what extent is it more sensible to focus on the relationships as such, to describe the things of the world as the sums of their relationships rather than highlighting their often-artificial separateness? The other category of questions has to do with intangibles in the world, the things that exist but lack physical forms. There are many of them, hauled out daily and put to frequent use. Without them we could hardly begin to understand the world; we certainly couldn't communicate with one another about it.

A common speculation over time has been that, if we could simply learn everything there was to learn about each atom in the universe, we could then predict the course of events going forward into the future. We could know all that would happen thereafter since all events involve the interactions of atoms and atoms adhere to predictable rules of physical behavior. Scientists studying subatomic particles have long since cast doubt on this claim, even understood as a thought experiment rather than any worldly possibility. Much physical motion is chaotic and only loosely predictable. At some scales it is not possible to learn both the location and motion of a particle. Further, it is not possible to study at this level (or perhaps any level) without altering the very system one is attempting to describe.

Even putting these issues aside, the claim that we might aptly describe the world in terms of its parts runs counter to much that we know about the ways the world works. Perhaps the physical world might indeed be *composed* only of particles; this is the intellectual stance known as physical or ontological reductionism (a being can be reduced completely to its physical parts). But this is far from saying that a thing can be *explained* in terms of its parts—explanatory reductionism—for the parts can and do interact in ways that give rise to traits and capabilities far different from those of the parts. As a simple example, hydrogen and oxygen—both gases at room temperature—come together to form water, which has traits and capabilities vastly different than those displayed by the elements separately. Moving to greater complexity, we might take a list of the chemical elements contained in a human body and obtain them from a chemical-supply store. Spread out on a table, the chemical elements would bear little functional resemblance to a breathing, conscious creature.

When parts come together to form systems they commonly exhibit emergent traits and capacities that are not displayed by the parts separately and that cannot—certainly by humans—be predicted by studying the parts. In many settings, the parts really make sense only when considered in terms of their relationships; they are what they are because of how they function in and as components of larger things. Significantly, the larger wholes routinely exert control over the parts in a causal sense. Consider: when a person consciously raises her arm in the air, she is, at the organizational level of the person, exerting control over the arm. As the arm moves the cells in it move; as the cells move their constituent molecules move. That motion, of course, is not predictable based on a study of the cells, molecules, and atoms in the arm alone. The motion is directed and caused at the higher level. The same can be said about systems in nature, about nutrient flows through biotic communities for example and about symbiotic relationships among organisms. Relationships count, not just in shaping the motions and functions of the parts as parts but by giving rise to patterns and capabilities not exhibited in or predictable from the parts as such. To look at a tallgrass prairie, for instance, displaying to the human eye perhaps 150 or 200 species of prairie grasses and forbs, is to see an organic system that is much more than the species as such. Physically the prairie might be reduced to these species of which it is composed (and the countless smaller organics, soils, waters, and so on), but it cannot be explained simply in terms of these parts, and much of the prairie, as a functioning community, would disappear by disaggregating the parts.

This perspective, it should be noted, is by no means a new one. Like atomism as a worldview it traces far back into history. Many natural philosophers have even taken the view that only larger wholes can be understood as such, never the parts in isolation. Indeed, a recurring view has been that only the world as a whole can be understood as a distinct entity, for everything else requires, to gain an understanding of it, a grasp of the ways it interacts with other parts. Often but not always this view has been accompanied by the sense that the world is guided and shaped by a divine intelligence or form of universal reason. This is not to say that parts do not exist as such, or that they cannot have value as such; they do and can. All species, humans included, use nature by detaching parts and consuming or otherwise using them. The concluding message is simply that parts are more than simply parts. They are constituent components of larger things, which are often disrupted if not destroyed when fragmented or disaggregated.

What is true of nature generally is also true in important ways of humans, understood not as organic systems themselves (although they are that, as we shall consider) but as parts of larger social and natural systems. Humans are social beings and evolved to live in social settings. A person can't be fully described as a functioning creature without being embedded in social relationships. It has only been in relatively modern times that people have been talked about chiefly as autonomous individuals—as freestanding parts, detached from social roles and cultural norms. Indeed, as sociologists have explained, a fundamental assumption of crisis-laden modernity is that the chief social unit of society is the individual as such, not the tribe, family, guild, village, or other group.

This emphasis on the individual has brought gains in the sense of recognizing and honoring individual worth. But it comes or can come at the expense of understanding humans as social creatures and how human flourishing is so largely dependent on healthy social relations. The emphasis has particularly brought confusion over the origins of moral value, human rights, and such institutions as private property. As will be considered later, all of these elements are social creations and are best understood as such (though as we will see, the facts of our existence play key causal roles). Private property, for instance, is inherently a system of social relations among people governing the use of things. As such it can only arise within a social setting and rest on social convention. Claims of rights are inherently claims made against other people within a social order. They have no meaning apart from such an order.

Tangible and Intangible

This mention of human rights links the issue of physical nature and emergent properties back to the category of issues having to do with the intangible stuff that also exists in the world. From an early point philosophers came to the conclusion that intangibles such as numbers, and basic logical and arithmetic relationships (two plus two equals four), must have some reality to them apart from human consciousness and social convention. Human language itself—our words, phrases, ideas—all have existence apart from any particular tangible form that they might take. Stories and fictional characters might also be said to exist in a meaningful way—we certainly talk about them that way—even though they too lack tangible existence. Beyond that, our brains give rise to images and imaginative visions and narratives that can enter our consciousness. They too are part of the world, and they don't lack existence merely because they would disappear upon the death of the last-living person who knew of them. Their disappearance when the last knowing person dies merely means that their existence on earth was temporally limited (as are all living creatures), not that they never really existed.

In making sense of intangible things—what are they, do they really exist—it needs noting that the line between the tangible and intangible is not easy to draw. For instance, how would we characterize relationships, patterns, or processes? They exist in nature as traits of physical things but seem to involve something other and different than the physical things themselves. Are they tangible or intangible? A pot of water is heated and begins to boil. Is the boiling itself—aside from the physical water molecules becoming gas—something that exists in the world? It is a physical process, sure enough, and we talk about it as such, as something that exists in a place and time. But boiling is a matter of physical change in things, and does not have independent physical existence. The same can be said of a family of people; we refer to it as a thing, yet it too has no tangible existence apart from the family members and their interactions. At some levels of organization we feel confident in saying that a thing exists: an individual human is a discrete thing. A person's constituent atoms have, by all accounts, formed a distinct, new thing that has real existence. But how much order do we need, how much novelty in the resulting emergent properties, before we're prepared to say that the parts have gone beyond complex interactions and improved functioning to give rise to a new thing?

This last issue might seem esoteric, but it plays a decidedly major role in our interactions with nature. We are culturally prone to want to see nature as parts, to see it in fragmented terms. We have words that refer to larger parts of nature—to rivers, forests, wetlands—but we're disinclined to see them as distinct organic wholes. If and when wholes do exist in nature, then we might be obligated somehow to respect them as such. At the least we might acknowledge what we have done when we disrupt or fragment them. No one would contend that murder is permissible simply because the corpse at the moment of death still contains all of its constituent physical elements. But what do we say when a rare biotic community is functionally disrupted so that it no longer exists as a community, even as its physical components are still around somewhere? Do we see the whole as such and consider it a distinct thing or do we overlook it and argue that nothing is lost when it disappears?

An important part of the cultural change now much needed is for us to see the world differently. We need to recognize larger wholes and key relationships as valuable, constituent parts of the world. We need to show greater respect for emergent properties. The overall need, in short, is to step back from our tendency to fragment the world, to see the parts without appreciating what they form. Ecologically and socially a wolf pack is more than the same wolves kept in separate pens. A heron rookery is more than a specific number of nests. Our tendencies to fragment might well be genetically in-bred. Hunter-gatherers were on the lookout for valuable parts of nature and alert to particular predators; only in quieter, spiritual moments did they have cause to reflect on larger wholes. Our world, though, is far different. Given our numbers, technology, and economy we need to pay more attention to patterns and connections, to ecological function, to emergent properties.

This leaves one final issue to take up in the foundational work of deciding what the world contains. The final issue has to do with a particularly vital intangible, the ideal or concept termed goodness or morality. Does it have independent existence apart from human consciousness and social choice? Plato famously thought that it did and vigorously pursued it. Many religious believers have thought the same, sensing in the world a moral order or moral power that bound and judged people without their consent. Enlightenment figures, as noted, tended to acknowledge the existence of a traditional moral order, as a real element of the universe, even as their tools for gaining knowledge couldn't detect its independent existence. By the turn of the twentieth century, the popular mind was finding it harder and harder to believe that a moral order was anything other than a social convention. It existed but as a human creation, and it was subject to change as human sentiment evolved.

In his important work on ethics, *After Virtue*, Alistair MacIntyre observes that people routinely talk about morality as having independent existence but have no intellectual way to ground it objectively. They treat it as real but have no way to justify the belief. Few philosophers see any way to fill this intellectual gap, not with just facts and reason. In any case, the wide variation in cultural practices among societies suggests that, if a real moral order does pervade the world, people have certainly struggled to find it and abide by it. Within religious communities, belief in a transcendent moral order typically rests on revealed truth and faith, not on some combination of reason and facts. Mac-Intyre has proposed an approach to morality grounded in solid virtues, which arise, he contends, from the facts of human nature and human potential rather than from social convention and human choice.

Prominent writers since the Second World War have also expressed similar support for an objective moral order —Richard Weaver in his postwar classic, *Ideas Have Consequences*; C. S. Lewis in, for instance, *The Abolition of Man*; Iris Murdoch in *The Sovereignty of the Good*; and Marilynne Robinson throughout her writings. But the academic consensus, without question, is that morality is inescapably based largely on shared human choice, even in the case of the bedrock moral claim that human life as such is morally worthy. Moral value can reflect physical reality as we experience it; it can align with our in-bred sentiments and evolutionary trajectory. But it does not exist in our consciousness and discussion unless humans proclaim it.

As we'll consider in chapter 3, human value judgments arise out of some swirling interaction of sentiment, facts, and reasoning, all influenced by genetic tendencies and cultural trajectories. There is, it would seem, no other source for morality, at least to which we have access. Among the implications: Just as long-familiar moral values have arisen and gained legitimacy in this way, so too can new moral values, more attuned to ecological facts and responsive to looming threats. We are value-creating beings, the preeminent ones on the planet, and the values we put forth and embrace are as solid and binding as any values can be.

Construction and Deconstruction

An appropriate ending point for this foundation-building effort is to take a look at one of the most confusing strands of nature-and-culture writing coming out of the academy in recent decades, the writing dealing with the "social construction" of nature. A brief consideration of this topic can accentuate and help bring together the key points of this chapter.

Central in social-construction writing is a claim, made in various forms, that much of what we perceive in nature and know about it is created within our minds. Nature to a large extent is a human product—so the claim goes—and we deceive ourselves when we keep talking about it as something that exists entirely apart from us. It is particularly confusing to talk about saving nature, or protecting it from people. If we've mostly created it, then what do saving and protecting mean?

There is, as noted, substantial truth to this reasoning to the extent it refers to limits on our ability to gain knowledge about the world and to the inevitable ways our minds filter sensations and construct images about the world. These epistemic limits, however, are best absorbed as cautionary measures to keep us from being overly confident in what we know. They do not suggest that nature itself is somehow artificial, that physical reality does not actually exist. That claim is simple nonsense. Our every action of living establishes the truth of our belief in nature's reality. Belief in nature's reality leads to far better results than would any belief that nature was a figment of human imagination.

A more plausible claim about social construction is that the patterns and organic wholes that we perceive in the world are based in part on human interpretation and choices generally. This too is certainly true, but our interpretations and choices doubtless have real facts behind them, including ample evidence of interdependencies and emergent properties. Some interpretations of nature are no doubt flawed; to the extent of their flaws they relate to a human misunderstanding—to a human creation—not to physical nature itself. It is true, also, that current interpretations might well give way to different, perhaps better ones as time goes by. But to recognize this human agency and fallibility is, again, not to doubt the reality of the physical stuff of nature. It is not to claim that organic connections and emergent traits are somehow imaginary. That stance clashes sharply with facts that are more than adequately proved.

What often gives this line of reasoning traction seems to be a failure to distinguish clearly between nature as physical reality and the language we use to talk about it. Language is a human creation, complexly aided by our genetic makeup. This is true of the words and phrases we use to talk not just about nature but about all other topics. And inevitably, words and descriptions are not the same as the things to which they refer. Perhaps in dealings with nature the need is greater than in other settings to keep this knowledge at hand, to keep aware, for instance, that the term "species" is a useful human-created category that organizes facts in the world but does not track reality exactly. (Each organism is genetically unique and lines between the categories we term species are commonly blurred by genetically intermediate organisms.) Still, to talk about the gap between language and reality is not to call into question that reality; it is merely to remind us of our challenges in learning about it and talking about it. If useful to us, there is no reason we cannot use the word "nature" to refer to nature apart from humans, even though we are embedded in it. This, it would seem, is no less legitimate than using the word "sky" to refer to atmosphere far above us even though that atmosphere reaches all the way to the ground, or to use the term "lake" to refer to a body of water that is, in fact, hydrologically connected with larger water flows and whose water content is constantly being exchanged.

Almost inevitably, it seems, commentary on social construction at some point turns to a consideration of wilderness. No component of nature has been more often labeled a human artifice. Understood as simply a word, wilderness is, like all words, a human construct, though to say that is to say nothing of particular significance. Its definition arises by social usage just as with other words, and different people use and have used the word in different ways. Though definitions vary, wilderness typically denotes in some way a natural area that is relatively unaffected by human activity. As for how much human alteration is too much, at what point a natural area is disqualified from wilderness designation, there is no clear answer. It is often said that the less change the better, but the point is contested (for instance, on the maintenance of hiking trails) and many wilderness protectors support human interventions to disturb natural areas when needed to mimic disturbance regimes that humans have likely disrupted. The earliest advocate for wilderness preservation in the United States, Aldo Leopold, contended that Forest Service lands might qualify as wilderness even though ranchers grazed cattle on them and continued to do so after wilderness designation. For Sigurd Olson, the lyrical wilderness advocate of the Northwoods writing in the 1960s and 1970s, wilderness was more an aesthetic quality. A trapper's cabin here and there or a bit of subsistence fishing and logging could add to the feel of wilderness rather than detract from it.

When wilderness is defined this way, as an area much less altered by humans than most, there are clearly places that meet the definition. There are, under such definitions, actual wilderness areas. We get a different answer when wilderness is defined as an area completely unaltered by people. By all accounts there are no such places anymore, if only because of human-influenced climate change, disruptions of stratospheric ozone, and the spreading of persistent pesticides such as DDT. (We have not, of course, studied all spots on Earth to test this point and such a study would itself bring at least some human change.) If this is so, then the ideal of a wilderness completely unaffected by humans is not just a human construct—true of all of our ideals—but an ideal that does not correspond with any physical place. It is an ideal against which all real places fall short. One might note, though, that this is pretty much the nature of many human-created ideals: they are set at levels beyond reach. No legal system reaches the ideal of true justice; no socioeconomic system provides for equality of opportunity; no culture is free of invidious discrimination. (We might also note that, given the vagaries of atomic motion, there is no physical example of a perfect circle.) As goals, such ideals can prove useful even when there are no real-world examples and no chance of achieving them.

Once we distinguish clearly between human language and physical reality the confusion on this point, on the alleged social construction of nature, rather quickly disappears. What remains, and solidly so, is a recognition that the natural world is exceedingly complex, far beyond our ability to understand it fully and to capture it in words and images. What remains too is the admission that, when we go about interacting with nature, we are in fact guided by language and images; necessarily we think with them and in terms of them. These traits are inescapable though we can considerably diminish their distortions. In the end, though, to admit this (as we must) is merely to define our earthly existence. It is to identify one of the reasons why we have struggled, and will continue to struggle, in our efforts to live in nature without degrading it.

TWO

Use and Abuse

The category of North American animals that has suffered the most acute decline at human hands in recent generations appears to be freshwater mussels, which once populated river bottoms in great abundance and variety. At the end of the nineteenth century Illinois fancied itself the button capital of the world due to the fertility of its mussel beds, especially in the Illinois and Mississippi rivers. Harvested there by the shipload the mussels were stamped into buttons to meet the needs of much of the world. Mussels, however, are largely stationary creatures. Once a mussel finds a place to live it secures itself there and pretty much stays put. Thereafter it is at the mercy of changes in the river, in the fluctuating water flow, in the chemicals the river contains and, most of all, in the silt load the river carries. Human uses of the rivers—particularly dredging to facilitate boat and barge traffic—unleash massive amounts of silt that clog, cover, and kill many mussel species. Uses of surrounding lands have also taken a heavy toll. Eroded soil runs off nearby fields. Artificial drainage efforts accelerate water flows during parts of the year, leading to deeper channels, streambank collapses, and water-quality declines. For many environmentalists the disappearance of mussels—many species of them exterminated, many others imperiled—is both troubling in and of itself and emblematic of the ecological deterioration of aquatic systems generally. The decline of mussels is a serious environmental problem.

This claim of deterioration, though, has not gone unchallenged. Interest groups that use the river intensively for travel or waste disposal sometimes question whether the loss of obscure mussel species really is much of a problem. The mussels serve no particular human need, certainly no need that is not easily met in other ways. Many mussels are adapted to live in peculiar aquatic settings that are disappearing and they are apparently not needed to perform any ecological function elsewhere. Attached to river bottoms, mussels go unseen by almost everyone. Indeed, few people know of them, which means few would miss them once they are gone. What exactly is there to worry about?

It is useful to take this question seriously, not just because it expresses a point of view held by many, but because it goes to the center of the whole issue of misusing nature. When we say people are degrading nature we judge their conduct negatively We are not merely describing action; we are evaluating it normatively and finding it wrong, immoral, or at least inexpedient, But on what basis is such a judgment made and who picks the standard to make it?

Defining the Problem

It is culturally revealing that most people, if asked, would have trouble composing a definition of an environmental problem. Nearly everyone could readily offer an example or two, a matter of pollution, usually, or toxic contamination. But it is harder to craft a categorical definition. It doesn't seem sensible to say that an environmental problem is simply a condition in nature we don't like or want—rain on a day we'd like to picnic, for instance. That seems more like an environmental condition that we work around. We could say that a shortage of water in the desert is an environmental problem, but that too seems more like a description of nature. We can lament the lack of water and want more of it; a water shortage can cause hardship, even death. Yet somehow that kind of situation seems distinguishable from the problems commonly viewed as environmental.

One way to define environmental problem, a definition with quite useful implications, is simply this: An environmental problem is a human activity that in some way involves the misuse of nature. This definition implicitly distinguishes between two types of natural conditions that we might dislike, one due to human activities, the other arising from the natural traits of a place. It reserves the term "environmental problem" for those conditions that humans have brought about, or more exactly it focuses attention on the human action that causes the condition and describes that behavior itself as the problem.

This definition helps in two ways. First, it draws attention to the human activity itself and makes clear that the adverse judgment implicit in the term "environmental problem" is a judgment on some human activity, not a complaint that nature itself is somehow second-rate. Nature is what it is and it operates in some ways rather than others. We can make the planet better from our perspective and sometimes have; we needn't accept nature as it is. But it helps to get clear in our minds that some of our changes to nature are bad rather than good, and we need a term that applies to such misdirected actions.

A second benefit of the definition is one that might seem like a deficiency, and that is the inherent ambiguity or even vacuity of the term "misuse." The term is plainly judgmental; it refers to a wrongful act. But without getting clear on the meaning of "misuse," the definition of environmental problem is half empty. Does it help to define an uncertain term by using another term, also uncertain?

The benefit that arises here comes precisely because the term "misuse" or "abuse" calls out for clarification. As it does so the definition highlights that we can't point a finger at some human-caused landscape change and slap on the label "environmental problem"—the loss of mussels in a river, for instance—without evaluating the change under a normative standard that distinguishes legitimate use from abuse. And we can't do that until we've crafted such a standard. Quickly, then, we get to the task at hand, to develop such a standard or test, a task that gains in complexity the more one engages it.

We can approach this point in a related way that highlights our rather poor collective efforts at crafting a usable test. We might imagine being invited to drive the roads in a local region and to study its land uses. Upon returning we are asked: Are the people of the region putting the lands to good use? A sensible answer would begin with a confession: the drive through the region has hardly yielded enough factual information to inform a judgment. Far more data would be needed simply to know what people were doing in the place, much less to trace the ecological, economic, and social implications. A sensible answer would also confess, though, that even possessed of full factual knowledge it isn't possible to evaluate the goodness of land use without a definition of goodness. Without one, facts are just facts. No amount of facts, no quantity of scientific data, can produce a judgment on land uses, good or otherwise. As already stressed, evaluation requires normative thinking. Fact collection and assessment, no matter how essential and carefully done, contain no normative content.

To live decently on land we must change the natural world, just as all other life forms change it. In the instance of the slow-moving rivers of the American Midwest, once so rich in mussels, probably no species altered them prior to the industrial age more than the beaver (*Castor canadensis*). This large rodent constructed so many dams on regional streams that something like 25 percent of all waterway miles in the Mississippi River basin were once backed up by their structural engineering. The resulting ecological consequences were immense. Habitat improved greatly for many species while taking a turn for the worse for many others. For beavers, the dams made the world better.

Our challenge in living on land rather evidently includes the need to come up with a sound way to distinguish use from abuse. Only with one can we judge whether a change to nature is an environmental ill. To date we haven't done particularly well at this foundational task. Indeed, although the challenge once thus posed is plainly important, it strikes most people as novel, a matter never before presented in quite this way.

Sustainability and Its Limits

Pressed for a standard of evaluation many people would talk in terms of "sustainability"—a popular term that thrives despite sharp criticism. We misuse land when our activities are not sustainable—that's the likely response. The term is helpful enough in that it draws attention to the long term and reflects concern for the plight of future generations. Beyond that, though, the term lacks much content. What exactly are we sustaining? Not, one hopes, longstanding modes of degradation. Many of the resources we use are nonrenewable in anything close to human timeframes; any consumption of them would seem unsustainable. Should we stop drawing down such resources given that we can't sustain any level of use forever? All human activities not only change nature but trigger ecological ripple effects that can spread widely. Those ecological changes, in turn, will require responsive adjustments in future human activities. Given ongoing ecological change we can't continue living precisely as we have in the past—we can't sustain our past activities—if only because the surrounding natural world is changing and we need to adapt to it.

Many proponents of sustainability have offered more precise definitions, making reasonably clear what needs sustaining. A problem with them is that they differ widely, so much so that sustainability—as one of its much-cited deficiencies—can seem to mean all things to all people. Better-considered definitions tend to be phrased not in terms of particular human activities directly, but instead in terms of the conditions of the surrounding natural world—its ecological functioning, quite often, or the types and numbers of wild species that inhabit an area. These are the things we want to sustain. But if we know the ecological terms that we want to respect and promote, then shouldn't the operative standard—the test used to distinguish use from abuse—be phrased in terms of those ecological conditions, rather than in the vague, incomplete language of sustainability?

More will be said about an overall normative standard, beginning in chapter 5. What needs clarity at this stage is how we have stumbled along for years with little direct thought about how we might distinguish use from abuse, filling the intellectual gap with the term "sustainability" (or a variant such as "sustainable development") that, when pressed, lacks much content when its varied usages are all considered.

Our poor performance at this line-drawing task is significant, for it is rooted in more than simple inattentiveness and bad politics. The failing is linked to our tendency to want science to answer policy questions for us and, going further, to a too-common misunderstanding about what science is and what it is not. It is rooted also in our tendency to think that questions of morality should largely be left to individuals to decide as they see fit so long as they respect the public order. In the case of rural land, for instance, we are prone to think that landowners themselves should decide how they will use their lands so long as they avoid overt harm to neighbors. The implication here is that land use is private rather than public business and that the morality of using nature—the moral challenge of distinguishing use from abuse—should be made at the individual level, not higher up. Finally and similarly, our failure to address this issue squarely reflects antigovernment sentiment mixed with longstanding liberal claims that government should remain neutral among competing visions of the good, allowing individuals maximum freedom to act as they would. These are powerful cultural leanings that go far toward explaining our collective failure to think clearly about the use-abuse line. They also help explain why so many people push away evidence of environmental ills, even when scientifically well attested, and why we resist even environmental reforms that would bring economic growth.

The Activist Impulse

This criticism of sloppy thinking is not limited to people who resist claims of environmental ills. It often seems characteristic, too, of people who pay attention to problems, and even to the subset of such people, activists, who step forward to help slow or deflect the downward slide. Here the problem perhaps comes simply from a sense that good intentions and a basic sense of direction provide enough guidance for the conservation work that needs doing; more serious thought, particularly cultural criticism, is unneeded. The problem also appears linked to a tendency to focus on rhetoric that makes sense to audiences where they are in the present, not rhetoric that pushes for fundamental long-term change.

Shallow thinking among nature's defenders too often shows up in their efforts to resist development projects that would significantly alter natural systems. A finger is pointed at an unwanted alteration—a shopping mall, a new road, a reservoir, or waste dump—and its expected ecological consequences are promptly criticized as hurtful. As for measuring the anticipated harm of a challenged project, the familiar practice is to begin with some natural condition as the starting point for comparison. Here is the anticipated human-caused change to the natural system, and all of the change will be harmful; that's the outgoing message, implied or expressed. Of course activists typically expect to bargain and accommodate, they expect some ecological change to take place. They view themselves, that is, as a needed counterbalance to powerful economic forces and the more effectively they push against them the more degradation they might contain. There can be, to be sure, good sense to this approach tactically and the motivation is both easy to appreciate and praiseworthy. But the beginning point is nonetheless often a blanket—if implicit—condemnation of all landscape change. The measuring standard, the beginning point, is a landscape (or a particular part of it) essentially unaltered by people. Alteration is inherently bad, and the less of it the better. When green advocates take this approach they too have avoided the hard work of distinguishing between use and abuse. Meanwhile, as they oppose intensive human land uses they open themselves to claims of misanthropy, of caring about nature and not about people.

In fairness, a firm-line strategic approach toward further invasions of natural areas can sometimes make considerable sense, particularly in landscapes that are already much altered. It can make particular sense when the conservation effort aims to preserve a rare piece of wildness in a landscape that is already being hard pressed to serve human needs. For critics of such nature protection, the antidevelopment approach can seem unduly stark and one-sided: The complaining green groups, it seems, simply want to preserve everything. But in significantly altered landscapes—that is, in most places where people live and work—such a firm-line policy appears one-sided (misanthropic) only when the assessment uses the current, altered state of the landscape as the point of beginning for tradeoffs. If instead the beginning point of assessment is set back in time, if the assessment includes landscape changes already made, then the bottom line can look far different. The proportion of land being protected by green interests becomes far less; the corresponding proportion devoted to direct human needs is far higher.

There are landscape settings, then, in which opposition to further change might well make good sense, places where any further development would go beyond proper use to become abusive. Even so, though, the environmental side in such clashes would do better to put forward a more thoughtful stance on how we ought to distinguish between the two. It would do better to articulate where it would draw the line between use and abuse. Lacking such a stance, the easy tendency for activists is, as noted, simply to label all change abusive. If all is abuse then humans have no rightful role on the planet and the fewer of them the better. No major environmental organization, we should note, takes this stance or anything remotely like it. But lacking a clear message on use and abuse their public stances are easy to misinterpret and their own judgments can get sloppy.

This brings us back to the continuing, severe decline of mussels in the usually slow-moving rivers of the American Midwest. Defenders of them would like to save all mussel species, ideally in something close to their original distributions and population numbers. They know full well that this cannot and will not happen. Their hope is thus an ideal from which compromises will be made—huge compromises in all likelihood, given the enormous political power of agriculture and other river users. But by taking this stance mussel-defenders are, in effect, using wilderness-like conditions as their measure of the proper use of nature, which is to say they would, by implication, limit humans to uses of nature that yield little or no ecological change. Such a policy stance doesn't really help citizens and communities think through the issues clearly and to generate visions of good land use at large spatial scales.

In the case of mussels on river bottoms, it might well be that humans could thrive on surrounding land while slowly returning mussel beds to ecological health (albeit without the species that are gone). Such an effort would require major changes in current land- and river-use activities, probably ending all commercial barge traffic on the rivers, removal of the many locks and dams, and significant reductions in the artificial drainage of surrounding lands. But these various changes might over time in fact yield net economic gains for the nation, not losses (as commonly assumed), given the high costs of maintaining the artificial waterways (dredging, levee building, and lock-and-dam maintenance), given the availability of alternatives in transport (rail), and given the sizeable economic benefits of healthy rivers. Most efforts to protect or enhance nature do bring about sizeable economic gains overall (albeit with winners and losers), sometimes sizeable enough so that the resulting environmental benefits end up costing nothing. Even in such instances, though, it clouds thinking to press wilderness-like conditions as the appropriate normative goal, or as the baseline for measuring abusive change, implying that all human alterations to nature are wrongful.

The call for preservation needs to be presented in other, more thoughtful terms. Preservation of part of a landscape needs to be situated intellectually and morally within a fuller vision of people living well on land in ways that can long endure.

To sum up, we have not done well thinking clearly about how best to distinguish use from abuse. This shortcoming by all appearances is widespread (though with scattered exceptions), including among many environmental activists. It is important not to push this criticism of the environmental effort too hard, for it applies only in selected settings. Many activists, for instance, labor to improve the healthfulness of food systems, fully agreeing that we need to use land intensively to produce food. In such settings, a wilderness-type vision plays no role. Other activists support sound forestry management that uses mixed-species, mixed-aged, selective-harvesting techniques to yield continuous flows of high-quality timber while also providing good forest habitat for wildlife; what they oppose is heavy-chemical monocultural tree farming. Still others labor to restore habitat for salmon and other fish precisely so that the fish stocks can once again become important wild food sources for people along the river. Nonetheless, the environmental movement as a whole has not presented to the public anything much like a clear definition of ecological degradation, tied to a vision of good land use at the landscape scale. It needs to do so.

The Call to Restore

This failure to think clearly about use and abuse also crops up in the work of ecological restoration. Restoration typically carries a positive connotation but it is an ideal or activity that is weakened by the same kind of vagueness as sustainability. In practice the restoration ideal refers not usually to a desired land-use endpoint but instead simply to a desire to move backward in time, a desire to undo changes that have already taken place. A historic building is restored when it is returned to some prior physical condition. Necessarily building restorers need to decide how far back in time to go: To the point when the building was first completed? To some later point of occupancy, after some change to the original building had taken place? Land restoration raises the same question. In the case of natural areas the challenge further increases because it is hard to know earlier landscape conditions with any precision (there are no blueprints available to consult). Natural-area restoration is more challenging also because nature itself continually evolves. Not all changes taking place in nature are human caused and it can be difficult to know which changes are due to human action and which would have occurred without human presence.

Aside from the question of timing—how far back to go in undoing human changes—there is an even more basic one: Why is restoration a good idea? What is its purpose?

Presumably the point of restoration is to reduce the ill effects of a human misuse of nature. But is the desired end goal a landscape that *is* being well used by people? Is the work, that is, guided by a distinct vision of use and abuse and by a plan to eliminate abuses while continuing or expanding the legitimate uses? This could be the case, and sometimes is. Quite often, though, restoration means pushing people off the land completely and returning it insofar as possible to conditions that preceded human arrival. The goal, that is, is wilderness-type conditions as of some point in time (for instance, before European settlers arrived in a place). Once again, the line between use and abuse disappears and all human change—or perhaps all change by non–native Indians—becomes abusive. Once again, the charge of misanthropy resonates, even as local communities are benefitted by the revived presence of nearby nature.

Many restoration advocates are likely to dissent from this claim, and with good reason. Some restoration does not aim to undo all human change. It looks backward in time but not all the way to a time before human settlement (or, again, before European-emigrant settlement). Other land restoration work seeks to cover over the ecological scars of misuse without any real thought of reviving the precise biotic community (in terms of species composition) that once existed in the place. Still, restoration as a call to action inevitably conveys a message of moving backward, not forward. It implies that human changes have been bad without any offsetting message that human changes to nature quite often are good. This omission can prove especially confusing and even hurtful for families that have used the land being restored— the farm field, for instance, that is being put back to native grasses. One message embedded in the call to grassland restoration is that the farming of the land was misdirected from the very beginning. It is not a message, needless to say, likely to go over well with families that have farmed in a region for generations and are proud of it, families whose ancestors may have labored to break the sod and drain the fields so that the land could produce food and fiber.

Restoration is clearly a much-needed enterprise today. It should be guided, however, like all environmental work should be guided, by a thoughtful vision of people living well in a place, by some overall understanding of what good land use entails at large spatial scales. If well conceived, that understanding would provide for lands and waters that are off-limits to intensive human use, places that promote components of good land use that do not directly serve the basic needs of people living today. Without such an overall normative vision, however, restoration can too readily also come across as misanthropic to those prone to question it. It can send the unhelpful message that all human change everywhere is bad and needs to be undone. More than that, it can suggest that environmental progress comes about by taking lands from human use and setting them aside, as if people could not live on lands in ways that sustained their ecological health.

Nature's Dynamism

The well-known reality that nature itself changes over time, that it is inherently dynamic, poses a special challenge not just for ecological restoration but for making sense of our place in nature generally and endeavoring to produce a vision of good land use. One problem linked to this dynamism is that nature seems to offer a moving rather than static target of healthy functioning and composition. What does it mean to respect nature when nature itself is changing? Much natural change comes about through the efforts of particular life forms, thereby posing a related question: If other species change nature and if their changes are legitimate, part of nature's dynamism, then how should we think about human-caused change? The challenge here gains depth when change is considered in evolutionary terms and is understood through a cultural lens schooled to interpret competition-driven evolution in progressive terms. If change comes about through competitive pressures that generate progress, how can that change be normatively wrong?

Like all living organisms humans are fully embedded in nature. The air we breathe, the water we drink, the food we eat, the sunlight that is so necessary for metabolism—all come to us from nature and we cannot live without them. Our psychological health is also linked to nature in complex, poorly understood ways. We evolved to live in particular types of places and can feel ill at ease when pressed or lured into alien living conditions. The parts of nature that we directly consume are, in turn, themselves also embedded in nature systems. The flourishing of terrestrial life forms begins with plants growing in soil and sunlight. The soil itself must be fertile, and its fertility depends upon complex cycles and flows of nutrients. Water is part of the productive equation and it must be available at the right times and quantities. Too much can halt growth, erode the soil's fertility, and bring to the land surface salts that diminish production. The interconnections and interdependencies are countless, quite often taking the form of symbiotic relationships in which particular organisms can endure only if other organisms also endure.

One force driving nature's dynamism is the complex set of mechanisms commonly termed "evolution." Individual species and mixes of species evolve in competitive settings, new species arising by gradual means and old ones declining and disappearing. Over time, evolution has led to the emergence of living creatures of extraordinary complexity. It pushes also toward natural communities that become more biologically complex over time as the number of different species rises and as the species collectively become more specialized and efficient in using the geophysical resources of a place. Collectively they do better at taking advantage of sunlight, engaging in primary production, and keeping nutrients in the local system as long as possible. These observable patterns, however, always have exceptions—systems in which fewer species dominate and localized species variety decreases; systems in which soil erodes rather than, as usual, slowly builds; systems in which primary productivity seems to decline rather than rise. Natural communities and thus their constituent elements are affected regularly by various disturbances, due to fire, flood, extreme weather, gradual climate change, and the like. All of them are forces of change. Vigorous biotic systems often recover from disturbances but never return to precisely where they were before the disturbance took place.

When Charles Darwin's writing on evolution first appeared, it came to a reading audience, particularly in Britain, that already believed in progress and already had faith in competition as a means of promoting growth. It believed also that large historical forces were at work in the world, biological and social forces akin to the physical laws that guided motion. It was easy for a readership like this to interpret Darwin's evolutionary theory of natural selection as just such a force bringing about progress. The stronger, faster, more physically able organisms were the ones who survived and reproduced. Their greater physical abilities, the mere fact of their competitive survival, meant that they were better than the genetic variations that lagged behind. Darwin, as already noted, did not frame his evolutionary theory as a story of progress; it was simply a tale of long-term physical change, he stated, aptly summed up in the neutral phrase "descent with modifications." It was Darwin's popularizers who gave evolution its overtly progressive caste and broadened its application into social and cultural spheres. Chief among them in Britain was Darwin's intellectual champion, T. H. Huxley, who had earlier come up with the phrase "survival of the fittest."

In educated minds of his day Darwin's writings on biological change were patched together with the many writings by other scientists on geologic shifts over time: on the creation of mountains and canyons, on changes in river courses, and on the various physical forces that, in dimly understood ways, somehow shaped physical landscapes over the eons. The physical earth was also evolving it seemed, perhaps also progressing, along with its many life forms. Further evidence of natural change soon came out of the new science of ecology, a science that studied organisms within their communities and biotic communities as such. Natural communities studied as wholes, it seemed, were also dynamic, most evidently when a community responded to a major disturbance by slowly returning to something close to its earlier biological condition. Change, in short, seemed to be part of the way the world worked. Other species, it was quite clear, altered the worlds in which they lived by ways of processes that at least carried suggestions of being progressive. Humans did just the same; they too were agents of natural change. By Darwin's day, few people outside religious settings believed that the earth had been specially created for humans. But few also seemed to doubt that it was entirely proper for humans to participate as agents in these ongoing processes of changing nature as they went about draining marshes, cutting forests, and redirecting rivers. Change was entirely natural and could be, it often was, good.

These various natural realities and cultural interpretations, all having to do with change in nature, have combined to make it more difficult for people to identify how they might best fit in the natural order over the long term. If nature were static and perfect, then the challenge would be to live on it and in it without materially disrupting its basic structures and modes of operation. But nature is dynamic. Its condition at any given time and place is contingent. As for the once-popular idea that nature is perfect, the notion makes sense only within a religious worldview that sees nature as God's handiwork and that attributes nature's perfection to God's own perfection. When that standard of evaluation is set aside, then nature simply exists and it operates dynamically in some ways rather than others. Nature is simply a matter of "is" rather than "ought." There is no right or wrong inherent in any natural conditions, no better or worse, except insofar as people evaluate it in those terms taking into account (as they likely would) their needs and otherwise using standards of their own choosing. As for nature's built-in dynamism, the changes it brings about are also neither good nor bad on their own. It is up to people to make sense of the changes, to view them through a normative lens.

In sum, we find ourselves needing to draw a line between the use of nature and the abuse of it and to do so in a natural world that changes on its own and that necessarily changes whenever we, or other species, live in it. Nature provides no static vision of ecological health for us to use as a guide. For that reason and for others, we cannot simply turn to any particular natural condition and say that the "is" of nature's existence provides a normative "ought" to use in judging how we interact with nature. (We return to the issue in the next chapter.) Nor can we sensibly use any sort of wilderness baseline as our measure of good land use, not for landscapes where we live and get our food. With God the creator largely put to the side—and, in any event, having abandoned any notion that God's perfection translates into nature's perfection—we are left to our own devices when it comes to distinguishing the good in nature from the bad. To be sure, nature works in some ways and not others, and we have abundant reasons to adapt our ways of living based on nature's functioning. Still, nature has hardly made our intellectual and moral work easy. It has simply not given us clear guidance on how we might successfully change it.

Ecology in Nature and Culture

Our best factual knowledge about natural systems comes to us from scientists. In the case of the biological functioning of natural systems it comes from the loose category of researchers known as ecologists. Ecology is one of the less solid sciences in that, in its study of the outdoors and its effort to make sense of it, the science necessarily deals with systems that it cannot fully control. Some ecology work is done in laboratories or by computer simulation. But most of it involves heading into the outdoors and gathering data from it. The data that await collection are essentially infinite—sobering reality number 1—and nature's parts and processes are all intricately interwoven—number 2. Inevitably, then, ecological study requires a scientist to formulate criteria for deciding what data will be collected from this infinite pool. Ecologists are good at this work, not just at developing data-collection criteria but also at formulating and undertaking experiments in nature to test hypotheses. Still, experiments conducted outdoors are never fully controlled, not like in chemistry laboratories. Conditions in different outdoor places are never the same, or the same in a single place at different times. The many factors that influence natural conditions challenge comprehension. Beyond that and beyond the already-mentioned limits on human cognition, there is the reality of human choice that goes into selecting the data and the necessarily tiny sample size that typically results. When studying a complex natural system, what features seem most important, and how can the countless features be brought together to describe the natural whole? A sense of importance is needed, which necessarily brings in values and culture.

A story told in recent decades about the history of ecology as a scientific discipline prominently features a claim that ecology underwent a paradigm shift sometime around or after the middle of the twentieth century. (Different versions of the story give different dates.) At some point, ecology as it summed up natural systems shifted from emphasizing the order and functioning of a natural system, presenting it as having rather stable, predictable elements, to a focus instead on a system's inherent dynamism and unpredictability. The old view tended to stress the ways in which natural systems endured over time. The new view tended to stress the ways in which it changed through random factors and from competition.

Such a shift certainly did take place, though it is a matter of opinion (and perhaps little consequence) whether it was material enough to qualify as a paradigm shift. As scientific knowledge has risen, there is greater awareness of the dynamic elements of systems, and greater recognition that systems, once disturbed, do not simply return to their earlier conditions. Community responses to disturbances are affected by more seemingly random factors so that a system, once recovered from perturbation, might differ considerably from its predisturbance state.

This new emphasis on community-level change, particularly when presented as the product of chance and competition, conveys a sense that nature really isn't a delicate, fine-tuned system that is easily degraded. Particularly if much of what we see in nature is the product of random factors—storms, the movements of seed carriers, particular rain patterns—then different random factors could have produced rather different systems, which we would view as equally natural. Maybe natural systems are tougher than we thought; maybe human-induced changes don't differ in kind from changes brought by other sources of disturbance and random processes. This message seemed implicit in the new ecology of disturbance, and it was a message that, predictably enough, was greeted warmly by industrial-interest groups when they got wind of it. The natural systems that people were disrupting were not just themselves dynamic, not just themselves products of actions and competition among life forms. They were the results of physical and random factors that could easily have given rise to something much different.

It is certainly true that this shift in ecology, however major it was, adds further intellectual complication to the line-drawing task while also suggesting that humans have greater freedom to manipulate nature than often claimed. Like all pertinent natural science, these new findings need to be given weight. As we do this, though, it is useful to keep in mind the points just made about the scientific field, both the exceptional challenges that ecologists face in making sense of nature and the roles of values and other social influences in shaping how ecologists go about their work. In significant ways, the overall shift that took place in the ways ecologists were talking about natural systems— the shift to emphasize change rather than enduring order—was rather directly linked to changes in the ways ecologists themselves chose to frame their inquiries and also in the motives that were driving their researches.

During much of the first half of the twentieth century ecologists talking about nature tended to consider nature's functioning over relatively short time periods, over decades or at most centuries (for instance, when describing vegetative succession). Their time horizon was relatively close. By late in the century, their successors were prone to talk more often about vastly longer periods—thousands or even millions of years. The longer time frames, of course, allowed for more change to unfold and their descriptions highlighted that more significant change. Earlier scientists often talked about systems apart from the natural forces that from time to time disturbed them. Their successors, in contrast, commonly found it more useful and accurate to include disturbance regimes as part of the system being studied, not external to it. This shift in focus, too, gave nature's dynamism a greater place in the descriptions they produced. Further, many scientists in the earlier period were interested in studying flows of nutrients and other components through natural systems, a type of work that paid less attention to the shifting species composition of a system over time. Later scientists, in contrast, often wanted to study these shifts in species composition and populations directly, which meant studying the components of a system that were often most likely to respond to random and competitive factors.

As a final point of difference, leading ecologists in the early to mid-twentieth century—Victor Shelford prominently among them—were often involved personally in the work of saving high-quality natural areas from degradation. As part of that work they needed to develop ways and terms to categorize natural areas by type so that samples of each could be protected for future study. (Shelford's Natural Areas Committee of the Ecological Society of America, an active agent for preservation, would evolve in time to become the US Nature Conservancy.) Necessarily this motive forced them to look at the community level and to describe it in a given point of time (typically the present), to come up with labels for types of communities, and to depict them in ways that seemed more enduring. Later ecologists, in contrast, left such work to conservation advocates. As they did so, they lost one of the main professional reasons for categorizing community types and labeling them.

Nature's communities, of course, did not literally come in discrete types; nor did they have anything like clear boundaries. Shelford and his colleagues knew that reality perfectly well, just as they knew that nature was dynamic, that systems once disturbed did not always return to original conditions, and that patterns of succession did not always lead to stable conditions. Yet their conservation work required them to talk about communities as distinct things, to situate them in time, and to give them a type of entity status. They also used this typology as a

61

means to introduce students to wide varieties of ecological communities in a type of education (learning thousands of species) that would disappear in the educational push for early student specialization. Looking on and then back, their ecological successors viewed much of this categorization and labeling work as artificial if not simply wrong scientifically in that it didn't correspond well with reality. Without the need to categorize lands for conservation, without any plan to take students (as Shelford did) on long summer trips to study wide varieties of natural areas, later scientists viewed this early work as unhelpful if not misguided even as it lived on in the work of state agencies charged with taking inventories of natural areas and identifying the best for protection.

These various differences over time in ecologists and their questions and motives surely offer a partial explanation for why late-century ecologists tended to stress change in nature more so than did earlier ones. Historian Donald Worster has added another interpretive component to the story by observing that scientists themselves, including ecologists, are never able to remove from their work their own values and understandings of the world (a commonplace view in the sociology of science). It was not coincidental, Worster concluded in his study, that ecologists late in the twentieth century tended to stress nature's dynamism right at the time when prevailing political and social thought in the Western World (the United States and Britain in particular) took a turn toward more individualistic, libertarian values, with a rising distrust of government and a tendency to see the world as a collection of competitive parts, not social wholes. Conservative values favored the free market—a dynamic, atomistic system—rather than the government with its top-down, centralized regulation. This view of the social and political realms, as chiefly individualistic and dynamic, rose largely in tandem with the ecological view that nature itself was in fact pretty much the same. Nature, too, was a chaotic system of individual competitive parts characterized by ceaseless change, just like the social order and the free market.

Whether or not Worster is right in his suggestion that culture and politics bled into ecology it remains true that ecologists struggle and will continue to struggle trying to make sense of complex natural systems. They will look at systems over time from different angles, asking different questions, using different time frames, and coming back with varied pictures depending upon the parts of the systems given interpretive primacy.

At the same time, however, the basic interconnections and interdependencies of nature are now quite firmly understood, including the processes that generate fertility and productivity. The basic ecological processes, including those responsible for primary productivity, are reasonably well described. So too is the fact that species come and go over time with new species very slowly displacing old ones and, to varying degrees, taking over their functional roles. In important ways natural systems do display distinct persistence over time—particularly in time frames relevant to human planning—and much change that does naturally occur is at far slower paces than the kinds of changes people make. Natural change that takes thousands or millions of years to unfold is, for human planning purposes, irrelevant.

Coming to Terms with Dynamism

These struggles by ecologists to make sense of the world further complicate our collective effort to know how to live. Still, ecological studies provide us the best knowledge we have and can get. We would be foolish not to make use of them even as we should not be surprised that new research both expands and modifies our understandings.

One particular danger arises out of this scientific knowledge and scientific uncertainty, and it is usefully kept in mind as we go forward. The danger is that we might look at nature's dynamism, particularly at the huge changes that unfold over thousands and millions of years, and view them as a green light to make big changes to nature of our own. Given that nature's systems are dynamic if indeed not transient and chaotic, then why can't we change nature as we see fit? Such long-term change appears to make ecological restoration seem unscientific in that restoration resists nature's own inherent dynamism. Dynamism can similarly make the preservation of natural areas also seem unscientific, in that even wilderness areas will evolve over time, gradually turning into something different. The intellectual risk here is hardly hypothetical. It is no surprise that, as noted, industry groups and neoliberal opponents of environmental limits have been quite quick to latch on to ecological writing that stresses change over stability.

Nature's dynamism needs to play a role in our normative thinking. That said, though, there are significant differences between the typical forms of human-caused change and those that take place due to nonhuman forces. Many of the differences have to do with matters of scale: natural change often occurs much more slowly and, in human time frames, is more localized. There is, of course, the occasional volcanic eruption that instantly alters hundreds of square miles of land. But changes of this type—again in human time frames—take place only on a tiny portion of the earth's surface, hardly appreciable when compared with the human footprint. In addition, nature's changes are caused by natural processes that have been in operation for eons, processes that various life forms have learned not just to deal with but often to turn to their advantage. Some human changes (plowing fields and exposing soil to annual weeds) are of this type while other changes (nuclear and toxic contamination, for instance) are not.

Perhaps the most important point to emphasize is this: Nature's dynamism is simply part of the "is" of nature. As such it has no more normative content than does any other trait or component of nature. The fact that nature acts in a certain way—that it changes over time— does not mean that the change is normatively good (or bad) from a human perspective. Nature's dynamism is hardly irrelevant; indeed it needs an important role in our standard for judging whether or not we are using nature legitimately. But natural dynamism does not provide an intellectual shortcut. It doesn't give cause for us to skip the hard work of thinking clearly about how we ought to live on land. Perhaps it gives us greater options to make changes without degrading nature. Perhaps instead it imposes greater limits on how we act if we want to maintain high natural productivity and to enjoy biological diverse surroundings. In any event, we still face the challenge of distinguishing use from abuse, the challenge of identifying the many relevant factors, of giving deep thought to our needs and aspirations, and of somehow bringing them all together into a vision of right living.

THREE

Science and Morals

An important tendency in the modern era, particularly in cultures shaped by individualism, is to turn to science to make sense of alleged environmental ills, to ask scientists to give guidance on them if not in a sense to arbitrate disputes about them. Journalists routinely seek out scientific experts to comment on possible dangers. Lawmakers do the same, not just when holding hearings but when specifying how regulatory agencies should implement broadly phrased statutes. Under the US Clean Air Act, for instance, the Environmental Protection Agency is told to set maximum air-pollution levels based strictly on science. Similarly, the US Fish and Wildlife Service, in deciding whether to extend legal protections to a species under the Endangered Species Act, is told to make its decision based entirely on science and other factual data. Science provides at once the vocabulary and the arena in which environmental policy often plays out.

This high status for science is consistent with a more encompassing cultural tendency on matters of public interest to embrace objectivity, to focus on facts and logical reasoning and keep emotions and personal preferences out of the picture. Scientists are viewed as the most objective and are raised up accordingly, even as they get attacked when the facts they generate are unwelcome. Indeed, the attack on science by defenders of the status quo—the denial of climate change at the moment most prominent— only highlights the importance of science; if critics can topple the scientific proof they might just carry the day.

Objectivity is deemed a virtue in the modern age, at

least in public affairs. It is not, of course, regarded the same in artistic and other expressive realms. Indeed, when it comes to personal spheres therapy is one of the age's leading tropes. The psychological professions, social work, personal counselors, even many churches: all are about helping people identify their subjective choices and become comfortable with them. Modern advertising, it hardly needs saying, is all about subjective yearnings and stimulating more of them. Objectivity still plays a role in personal matters; there's little room for subjective expression when fixing a leaky water pipe. Nonetheless, the contrast between public and private spheres is rather stark. Objectivity dominates (or is supposed to) in the public arena. Subjective choices are given greater rein on the private side.

The still-lingering debate about human-induced climate change offers a case in point. Three elements of this debate are particularly illuminating in cultural terms. First, scientists are put front and center in it, both the thousands of scientists assisting or agreeing with the Intergovernmental Panel on Climate Change (IPCC) and, disproportionately, the vastly fewer scientists who question the dominant consensus. Atmospheric scientists are expected to tell us whether our modes of living are problematic. Second, the issue as framed publicly is whether or not human-induced climate change has been proved as a matter of fact. As raised, the issue calls for factual evidence, collected and weighed. Finally, when it comes to proof the preferred standard is that of scientific proof. Has it been factually proven, in the scientific sense, that humans are materially changing the climate?—that's the question.

On all three of these points, today's climate-change debate reflects distinctive cultural traits, ones that are, in this setting and others, rather confusing and unhelpful. Indeed, most of the reasons why the modern age has trouble coming to terms with ecological change can be teased out of this slanted, three-part framing of the climate issue. To see this, though, we need to back up. We need to explore what science is and what it is not. We also need to consider the origins of morality, or more generally the origins of normative values—the values or standards used to distinguish the wise from the foolish, the ethical right from wrong. Where do we get the raw materials to make such determinations, and what gives them legitimacy? Both of these inquiries—on science and morality—lead to rather firm ending points in that the foundations of both are reasonably clear. We need to gain better awareness of these foundations if we are to make sense of our ecological plight.

Objectivity and Its Costs

When people talk about science, scientists included, they typically have one or both of two meanings in mind. Nonscientists typically refer to the body of factual knowledge that has arisen from or been confirmed through the scientific processes. Science is what we know pretty much for sure. In the case of natural science it is what we know about the natural world, its constituent elements, how it functions, and how it changes over time. Scientists use the term this way but they also use it to refer to the methods and standards by which new knowledge is generated and tested. Science is a process, guided by professional standards. It entails formulating and testing hypotheses and gathering and interpreting data under circumstances that are sometimes controlled, sometimes not.

What is important about science as thus defined is that it is all about facts and their interpretation (setting to one side science as engineering and technology). Science as method or process is a purely descriptive enterprise in the sense that it seeks understanding about what is, what was, and what will be. The sought-for descriptions can cover dynamic processes, not just static conditions. They can look back in time and forward, and include predictions based on stated assumptions. What they cannot do is pass judgment on the goodness or badness of any particular state of affairs, not without drawing upon at least a vague normative standard pulled from elsewhere.

We might consider, for instance, how scientists would describe two equal-sized fields, one covered by a tallgrass prairie of the type that once dominated east-central parts of the United States, the other a familiar expanse of soybeans planted in rows. Scientific descriptions of these fields would differ substantially in terms of their resident species (macroscopic and microscopic), the functioning and interactions of the species, nutrient flows, hydrology, and more. The descriptions would vary in both static and dynamic terms. Yet, the scientists doing this work could not, if pressed, tell us whether one field was better than another, or whether the condition of one field was more morally right, beautiful, or even useful. These questions, as noted, can't be answered without drawing on standards of evaluation that come from outside science. The scientific data are, of course, essential to any evaluation; normative standards alone are hardly enough. Answers require that the two parts be brought together. Scientists might be the best people to do this work; they'll have a better grasp of the often-complicated scientific facts. But it is work that reaches beyond science as such, and the conclusions of any assessment, whoever does it, are sound only insofar as the right evaluative standard is used along with good facts.

To see this is to see why it is problematic for atmospheric scientists to be expected to explain whether human-induced climate change is worrisome. The science part they can take on, challenging though it is. But the judgment about whether ongoing change is problematic requires use of a standard of evaluation. What is the best one to use and who gets to pick it? We could, of course, view any human-caused climate change as stupid or immoral, embracing a zero-tolerance policy. But to take this strict approach is, once again, to turn against the idea that people belong on the planet and can legitimately use it. It is to assume that all change is abusive and the less of it the better. Perhaps such a strict standard does make sense when it comes to climate; our knowledge of climate change is distinctly partial, and we don't understand in particular how climate change, once it gets going, can feed on itself. But the absolute, no-change-to-nature standard is typically unhelpful in dealings with nature. It is not immediately apparent why it would make good sense in this setting.

Our tendency to treat climate change this way, treating the issue as one of science and expecting scientists to give advice, says much about where we are culturally. The prospect of human-caused climate change triggers profound moral concerns. Our tendency, though, is to view moral questions as properly lodged in the personal sphere of life, not as matters for public judgment—at least in the case of issues such as climate change that seem to pose new quandaries. Former vice president Dick Cheney expressed this perspective when he defended a new energy policy that only considered ways to increase energy supplies. The policy didn't consider energy conservation, he asserted, because conservation was a personal virtue, not a matter of public policy. No doubt Cheney's oil-industry ties had something to do with his stance. But his reasoning likely resonated with many people. Morality was like religion, a matter for people to sort out and implement in private life so long as they didn't harm anyone else. Subject to a no-harm rule and with due regard for the equal rights of others, individuals can make their own subjective choices.

With this cultural slant, society as such has real trouble framing the climate-change issue sensibly. When morality is mostly about one-on-one interactions and individual rights, how can we talk about what is good for us collectively? How can we talk about moral obligations that we should bear not as individuals but as a people acting together? As taken up in the next chapter, public debate does often focus on economic growth, viewed as a desirable (normative) collective goal. And depending on how growth is measured, that goal can incorporate normative elements linked to collective welfare. But as we shall see, economics at root also places emphasis at the individual level, on the preferences people have as individuals. It is no substitute for direct public engagement on the wisdom or folly, the rightness and wrongness, of particular public policies. Our willingness to talk (to perseverate really) about economic growth doesn't deviate much from the broader tendency to leave moral issues for individuals alone to resolve.

This public cult of objectivity is by no means normatively neutral. At first glance it may appear so, that a government that leaves people free to make their own normative choices is simply acting as referee, keeping the peace but not taking sides. But even a bit of probing disproves this claim. For instance, a government that takes no stance on abortion in effect allows it to take place. Permitting abortion is no more morally neutral than banning it. In the same way, a property-rights regime that allows landowners to destroy critical wildlife habitat on their lands is no more neutral on the matter of species protection than a system that obligates landowners to protect habitat. In both settings, public inaction is a value-laden choice. In both settings, the moral issue is resolved, not by addressing it directly at the communal level, but by turning it over for individuals to address separately. This route might aid public harmony. It might promote human welfare by empowering individuals to assert control over their lives, their bodies, and their immediate surroundings. But it is by no means normatively neutral on an issue where one person's choice affects other people or the community as such.

When this happens, when an important policy issue is simply turned over to individuals for their free choice—based (for instance) on some strong preference for maximum individual liberty—then the normative issue too often is never really engaged and fleshed out in the public sphere. What gets missed, what often goes unidentified, is the initial, normatively charged question: whether decision-making on the topic should occur at the public level or whether instead it should be reserved for individual choice.

The strong push to protect if not enhance individual liberty, so pronounced in the United States and increasingly so elsewhere, is thus a political stance with considerable implications. This is particularly so when it comes to dealings with nature. Many normative elements of right living in nature can only be achieved by people acting together—

protecting rivers as systems, for instance, and migratory wildlife. They require implementation by government, acting as agent of citizens. To push decision-making down to the individual level in such settings is effectively to rule out critical options, often without realizing it. The bias of the approach is toward options that people can act upon individually. That typically means slanting choices toward courses of action in which the resulting benefits are ones that individual actors can enjoy personally, pretty much without sharing with others. Why pursue a costly course of action when the benefits go in significant part—maybe almost entirely—to other people unless the other people are obligated to act the same?

The Limits on Proof

The piece of the climate-change story not yet covered has to do with the issue of proof. The modern tendency, as observed, is for scientists to be asked for answers and the question posed thus is factual: Are we in fact changing the climate in a significant way? Merely to pose the question to scientists is implicitly to ask for an answer using scientific methodology. Has it been scientifically proved that we are changing the climate?

Countless environmental issues are reduced to this same question. It is a common practice, one that needs unpacking to expose its cultural meanings and social consequences. What does scientific proof mean, is use of the standard sensible, and what does it say about modernity that we so regularly and instinctively employ it?

The issue of proof is linked to the definition of truth, a matter taken up in the first chapter. For scientists, the gold standard, the aim of all inquiry, is to establish truth using the correspondence definition. A statement about the physical world is true if it corresponds accurately with the physical conditions of the world, if it accurately describes reality. To the nonscientist, the standard seems plain enough and easily satisfied many times. But scientists bring more critical judgment to bear. Philosophers of science are even more demanding, so much so as to assert that no facts are every fully proved, particularly facts about nature. Instead they are simply established to very high degrees of probability, always leaving open the possibility that new data will call for modification.

As already noted, explanations of scientific proof typically distinguish between two modes of reasoning, deduction and induction. Deductive reasoning begins with initial axioms, presumed to be true, and proceeds by logical steps to conclusions that are, in a sense, implicit in the axioms. Mathematical reasoning is of this type. When the logic is sound the conclusions reached are said to be proved, though of course their accuracy depends on the validity of the starting axioms. In the study of nature, including the study of ills such as climate change, deductive proof plays a distinctly secondary role.

More central to factual claims about nature is inductive reasoning, which begins by gathering sensory data about the world and inferring conclusions from them. For instance, if numerous balls are dropped and they consistently fall toward earth at the same rate, a conclusion can be drawn about what will happen when the next ball is dropped. Data about the falling balls are brought together and an inference drawn that the next ball will move in the same way. This reasoning seems sound enough and very likely is. But it depends upon a key assumption, first prominently explained in the eighteenth century and accepted ever since. The reasoning depends upon what is termed the Uniformity of Nature, the assumption that nature exists and behaves in uniform ways. Without that assumption, an inference about the next ball's motion is not logically sound. So far as scientists can tell, nature does act uniformly in many respects but patterns in nature are by no means always uniform. Particularly when it comes to actions involving living creatures they can vary in ways we might easily miss. On the other side, nature might be acting uniformly without our knowing it because it follows a pattern more complex than we have observed. Nature's dynamism comes into play, so a pattern that prevails for a time may also shift due to natural causes.

The illustration of the falling ball is a simple one in that experiments are easily undertaken under controlled circumstances. Balls can be dropped thousands of times, just as coins can be tossed thousands of times, with the results recorded and tabulated. In many settings, however, experiments are costly and time-consuming if they can be undertaken at all, and they may take place under conditions that are erratic. Often data must be collected simply by observing events as they unfold outdoors with no ability to control the operative forces.

Here we might consider a study of forest logging and its effects on the nesting success of various bird species. When a solid block of forest is disrupted by the logging of wooded patches here and there, what happens to the ability of forest-dwelling birds to raise their young? Realistically scientists can't take control of multiple forests and conduct experiments in them again and again. Even if they could, the forests would differ in species composition, climate, hydrologic flows, and more. Inevitably researchers must gather real-world facts as they can and infer conclusions from the facts. To the extent the data show material consequences from the logging, the researchers then must formulate theories to explain them. As more studies are conducted, necessarily under varied conditions, more data come together and explanatory inferences (often revised) can gain strength. But scientific conclusions in such a setting are never as solid as in the case of balls being dropped. This is so because the assumption about the Uniformity of Nature seems less secure. It is so also because study conditions are uncontrolled, the likely relevant facts are vast, and the possible explanatory factors and forces so numerous, more numerous than the simple process of gravity at work on the falling balls. As a result, conclusions necessarily are more tentative. In the language of the philosophy of science, a conclusion in such a research setting is often termed an Inference to the Best Explanation. Scientific judgment is needed to decide how solid an explanation seems to be. In any event, a conclusion cannot be proved in any airtight sense. New data might call the inference into question and future scientists, reviewing the same data, might propose different theories of causation.

Given these limits faced by scientists, it is often said that science cannot prove claims to 100 percent certainty. Instead it can offer claims with varied levels of confidence behind them, based on probabilities that can approach 100 percent but never reach it. The point has been known for centuries, and it formed a core element of American Pragmatism at the turn of the twentieth century. Pragmatists added to this insight when they put forth the third of the definitions of truth by which the truth of a particular claim or proposition was characterized by widespread acceptance of the claim by scientists in a field and by the consequences that flowed from accepting it and taking action based on it. As the prominent philosopher of science Karl Popper forcefully explained in the twentieth century, science had the power to disprove propositions conclusively but could never prove anything completely. It had to get by with elements of uncertainty that could and did vary considerably among scientific disciplines based on the challenges they each faced and the limits on available research methods.

These observations about scientific proof are especially pertinent to the case of long-term climate change. Plainly, atmospheric scientists cannot conduct whole-planet experiments in which they study the effects of human activities over very long periods of time. We have only one planet, people are living on it and changing it, and nature's ongoing processes are dynamic. We can't put the planet on hold while we conduct thousand-year experiments. Further, the relevant data awaiting collection are essentially infinite and scientists must get by with well-chosen pieces. Their goal can only be to draw conclusions by inference both on what the facts are—whether change is taking place and if so how much—and on why the change is taking place. Necessarily any conclusion would be a matter of probability. Necessarily also the accuracy of conclusions will be evidenced in the same way as other scientific truths: by widespread acceptance among climate experts and by the good consequences that come by assuming their validity (in the form, for instance, of newly collected data that conform to predictions). Scientists working in and with the IPCC know all of this full well. From the beginning they have presented their factual conclusions only as matters of probability. Over time, with each new multiyear assessment, the IPCC's conclusions have typically (although not on all points) been set forth with higher levels of probability. Sadly, journalistic summaries of their work often omit the words of probability, or mention them quickly and then push them aside, with rarely an insightful comment on the methods and constraints in science of this type.

Public uncertainty about scientific proof has hampered the public's understanding of climate change. The poor understanding opens the doors to demagogues who assert that if climate change were true all data would support it and the proof would be certain. More troubling than that is the fact that the burden of proof being publicly used is a scientific one. Scientific standards play essential roles in the scientific process. But is it wise to use them outside of that arena? Is it wise to insist that facts be accepted in public affairs only when established to an extremely high confidence level?

Other Burdens and Their Values

What is commonly termed "scientific proof" is only one of many burdens of proof in regular daily use. Burdens of proof are the stuff of law practice and legal systems. In civil courtrooms in the United States, facts are accepted as proved if they are supported by the preponderance of evidence adduced at a trial, which is to say supported by 51 percent of the evidence. The standard is higher in the case of criminal trials. There, the prosecution is obligated to prove key facts beyond a reasonable doubt, a standard that defies translation into numerical terms but is certainly below 100 percent. In other legal settings different standards of proof are used. An intermediate one between these two is proof by "clear and convincing evidence." An even lesser standard is one in which factual conclusions are treated as adequately supported unless the underlying evidence is so insubstantial that the conclusions seem not just unlikely but arbitrary and capricious. No judicial proceeding requires that facts be supported to the level of scientific proof. Indeed, criminal defendants in the United States are put to death on lesser proof than that.

For further comparison we can turn to daily life. We routinely exercise caution to avoid dangers that are unlikely to happen. Often we are unwilling to assume even small risks of harm. Who would get on an airplane, for instance, if told there was a 50 percent chance of the plane crashing, or even a 5 percent chance? It is hardly sensible to ignore such a danger. It is hardly sensible to brush it aside on the ground that the factual prediction of an upcoming crash has not been scientifically proven. Who would eat food that was 10 percent likely to cause serious illness? As the extreme of caution we might consider the case of the US Secret Service, charged with protecting the president. The Secret Service, we can presume, takes a death threat seriously and acts on it even if its chance of happening is 1 in 1,000, or 1 in 100,000.

In this light we can reconsider our social tendency to frame climate change in terms of scientific proof. Is it not more ethical and sane to pose a much different question: Is the evidence in hand indicating that a problem looms ample enough to merit a remedial response? If that were the question, how much evidence of possible harm would we require? How likely would the danger need to be to prompt corrective action? Presumably our answer would take into account the costliness of the correction, assuming (as we do) that actions to reduce climate change would entail net costs, that is, costs greater than the non–climate related benefits they would also generate. (As an aside, we can note how critics of climate change, insisting on yet higher levels of proof of harm, are at the same time often inclined to grab tight to highly speculative claims about the net costs of halting fossil-fuel usage.) Aside from the cost issue, important normative factors are also highly pertinent when asking how much danger is too much—factors of morality, social justice, and the wisdom of precaution.

At root, it is hardly possible to defend the use of scientific proof as the appropriate burden when talking about potentially catastrophic harm. A much lower burden seems manifestly in order. So why do we still talk about scientific proof? How did the public issue get framed like this to begin with?

To the extent this burden was pressed upon society by climate skeptics—mostly industry funded—it amounted to an extraordinary rhetorical victory. The use of the scientific standard strongly skews debate to favor inaction and the status quo. It also means a scientist hired by critics can publicly claim that climate change has not been proved in the scientific sense while privately admitting that it was in fact highly likely to occur. The strategic value of this mode of resistance, centered on burden of proof, has been known for generations. In her bestseller *Silent Spring*, from 1962, Rachel Carson made the case that we should, as a matter of prudence, exercise greater restraint when using deadly pesticides given the ample evidence of their dangers, evidence that Carson reviewed at length. (She also contended that the use of pesticides, directly on people on their lands, violated individual rights since people had not given informed consent.) Carson's industry-funded critics, however, immediately pushed forward a different frame for the debate. In their responses, they contended that she had not scientifically proven all of her allegations about pesticides. The issue, they replied, was one of scientific proof. Carson's language of prudence and caution (and rights violations) got shoved to the side.

In terms of climate change, it is revealing and dismaying that public discussion (particularly in the United States) shows little awareness of the vital normative considerations that go into selecting a burden of proof. This lack of awareness, though, is consistent with the many other ways in which normative issues are pushed out of the public arena, usually into the private realm or, as in this instance, off the table completely. It is a significant intellectual failing. It compromises our collective ability to make sense of climate change and explains, better than any other factor, why the reality of climate change remains contentious. Had the facts of climate been submitted to a criminal jury, the kind that hands out death penalties, the claim of climate change would long ago have been proved beyond a reasonable doubt. A civil jury using a preponderance of the evidence standard would have drawn the conclusion a generation ago.

Before turning from science to the foundations of morality, two further points might usefully be made about our cult of objectivity and exaltation of science. The first has to do with the longstanding confusion about safety and what it means for something to be safe. (The same confusion surrounds the word "risk.") What does it mean to say something is safe—genetically modified food, for instance, or hormones or antibiotics fed daily to animals destined for human consumption? As widely used the term has several quite different meanings. Debates over safety are often confused and degraded by a failure to get clear on them.

Safe can mean a zero risk of any bad consequence. Using that definition almost nothing is safe. It isn't safe to get out of the bed in the morning, nor is it safe to stay in bed. It certainly isn't safe to drive a car or ride a bicycle. Safe can also mean that the risks associated with something are so minor or trivial that they can be ignored. Beyond that, it can mean that the benefits associated with something are more significant than the expected harms so that there is, on balance, a net gain. As should be apparent, the second and third of these definitions require an assessment and weighing of dangers and costs, or an assessment of both benefits and costs—an assessment that entails, again, the use of normative standards. Safety in the second and third senses is thus not simply a matter of fact. It is not a matter upon which science alone can pass judgment. It is perhaps plausible enough for a Monsanto to contend that its genetically modified crops are safe, but in fairness it needs to release all of the evidence that it has used in making the assessment and also explain clearly the normative standard (including the burden of proof) that it has employed.

Second, the push to have science as our public arbiter is sometimes associated with an insistence that the only evidence relevant to an inquiry—the only evidence that can be taken into account when assessing the truth of a claim—is evidence that takes the form of scientific studies in peer-reviewed journals. On its face this insistence would seem to improve the accuracy of ultimate findings. Sometimes it does. But it will do so only when and if peer-reviewed studies cover the relevant data with considerable thoroughness. When that is not the case, when published studies have not (yet) taken into account much relevant data, when they cover only scattered or spotty aspects of an issue, then conclusions from peer-reviewed studies might rightly and usefully be supplemented.

Here again we can consider the courtroom, where evidence is admitted so long as relevant and not plainly unreliable or prejudicial. There is no insistence that evidence be limited to scientific studies. Similarly, in everyday life risks are immediately processed based on experience and judgment. In the case of climate change, for instance, we have reams and years of recorded testimony by birdwatchers about how migration patterns of many bird species have shifted in ways consistent with a changing climate. Evidence of this sort, though, is criticized as anecdotal and unreliable. Climate critics insist that it be ignored. But evidence of this sort from hundreds or thousands of sources has much value. As important, it is the only kind of evidence that ordinary concerned citizens can contribute to the public discussion. To insist on using only peer-reviewed studies is not just to push aside valuable, pertinent evidence but also to push aside ordinary citizens and keep them from playing a role.

Reason's Attack on Morals

For many centuries people generally, Western philosophers included, largely took for granted that the world was structured or guided so that particular actions could be objectively right or wrong, or good or bad. (Some prominent ancient philosophers had said otherwise.) Actions or states of affairs might be wrong or bad because they clashed with some transcendent ideal of goodness and justice. They might instead be inconsistent with the purposes or end goals immanent in a creature or thing. They might disobey the wishes of a god or spirit or conflict with revealed religious wisdom. The variations were many, particularly when the known world included creatures and particular places that were inhabited by potent and demanding spirits.

To trace how we got to where we are today, in the ways we think about morality and its origins, we can take up the storyline with William of Ockham, a fourteenth-century English friar who as a Franciscan worked with people and nature outside the cloisters. Ockham became an early, prominent critic of the prevailing belief that ideals such as goodness and justice had real existence, that they were universals or Forms, in the language of Plato. To the contrary, Ockham contended, ideals such as these existed only within human minds as mental concepts. They were words that we used to express things within our consciousness, not labels linked to real things that existed apart from us.

In Ockham's day morality was chiefly the province of religion, based on instruction from God. It was thus significant when Ockham asserted that it was not possible—either by studying God's creation and gathering sense impressions from it, or by reasoning from those sense impressions—to draw conclusions about God or about religious matters. One could only know about God by means of revelation, Ockham insisted, by way of scripture or through direct spiritual insight. This division of sources of knowledge in effect separated the world into two realities: the reality of the empirical world given by direct experience, and the reality of God and God's teachings known only through revelation. Put simply, there was religious truth and scientific truth, and the two were not directly linked.

Ockham's reasoning had many consequences. In time his view helped free scientific inquiry of nature from the strictures of the church. If nothing learned from the study of nature told us anything about God, then no scientific conclusions could be viewed as blasphemous. As important, this separation detached data collection and inductive reasoning from any inquiry into the moral order. To grasp morality, some other means of inquiry was needed.

Ockham's two-part division of reality and knowledge worked fine so long as religious faith held up and knowledge by revelation was deemed as sturdy as empirical knowledge (as he believed). But with the continued advance of scientific thinking the two realms of reality became increasingly less equal, particularly by the seventeenth century with its rising commitment to empiricism and induction. By the following century, the Enlightenment Era, faith in man's powers and knowledge had risen high while scriptures and other revelation seemed less and less believable, in part because they lacked support from the senses and science. As science continued to rise, it became ever harder to keep the two forms of reality separate. Step by step Enlightenment leaders challenged revelation, moving further down the road to secularization. Yet, as historian Carl Becker pointed out in his influential study, *The Heavenly City of the Eighteenth-Century Philosophers*, Enlightenment figures nonetheless mostly retained a firm commitment to Christian morality. They continued to anticipate movement toward some form of heaven on earth. Thus, even as they pushed aside the Nicene and Apostle's creeds (and on to deism in some cases) they retained faith in an overriding order, structured by both moral and physical laws and, *contra* Ockham, discoverable by human reason. With God collapsed into nature one could, by studying nature, gain wisdom about the transcendent moral order and how people were supposed to live.

The trouble with this stance was soon evident. The practical study of nature and the use of reason simply didn't yield much in the way of moral instruction, at least much that was unambiguous. As early as the mid-century, Becker concluded, leading thinkers were admitting the feebleness of human reason on moral issues and were softening their caustic attacks on tradition and church. Nature and Nature's god didn't seem to have much guidance to offer. Various Scottish philosophers were among the first to shift ground and propose a new foundation for morality. Our moral senses, they asserted, arose not within our rational minds as long believed but instead from the emotions or sentiments we experienced as we engaged with the world. By this they meant not transient feelings but our more deep-seated, long-term sentiments about right and wrong, sentiments that, they thought (or at least hoped), were strong and stable enough to support something close to real, binding moral standards.

Such sentiment-based claims were listened to attentively even as the defenders of science and reason still believed that they held the keys to progress. Before the century had ended, however, another disruptive element had gained enough power to unsettle moral thought. This was the rise of liberal individualism, the rising belief (fueled by the late-century revolutions) that individuals as such were not just morally worthy in the eyes of God but were endowed in some manner with individual rights.

The idea of individual moral value, shared equally by all people, had become a central element of Christian thought in the Middle Ages, even as economic and social orders remained highly stratified. By stages the reasoning gained ground, breaking forth initially with claims of religiously grounded natural rights and then with rights that secular realms needed to respect. Public morality, it was claimed, had much to do with the recognition of and respect for these rights. It had to do also with the rule of law, similarly gaining strength. Moral thought should take the form of rules that bind and apply equally to all people, so said the late century's leading philosopher, Immanuel Kant. And those rules, Kant contended, should be ones that respect the worth of individuals as such, rules that treat each person as a morally worthy subject rather than merely as a tool or object. Moral thinking, then, properly begins with the individual human considered in isolation, not with the overall natural world and with an effort to understand its inherent order and how people rightfully fit into it.

This new moral reasoning in retrospect was based on human egoism, even as early adherents (Kant included) tended to retain the moral principles of Christianity (Lutheranism, in Kant's case). Kant framed his moral reasoning in terms of the moral duties borne by individuals; it was duty-based reasoning. Soon this reasoning came to be thought about as rights-based given that an individual's duties chiefly related to moral obligations, first, to respect other people as morally worthy subjects, not as mere objects, and, second, to live according to moral rules that one would want to apply equally to all other people.

Kant's moral reasoning spread widely. In time it would become one of the two dominant forms of Western moral thought, referred to as the deontological (duty-based) approach. It contrasted with modes of reasoning that judged moral rightness and wrongness based on the consequences of an action, particularly the effects on human happiness or flourishing. Kant's new reasoning, though, was not immune to the problems that Hume had earlier identified. Kant had to assume that humans were morally worthy subjects, just as Christian tradition taught. It was an axiom that he took to be, in Jefferson's terms, self-evident, not one drawn from facts or pure reason. Yet if humans were worthy, why not other creatures as well, and why didn't moral worth reside in families or villages or tribes along with, or instead of, individuals? Neither facts nor reason could explain why one starting point was sounder than another. Further, Kant similarly carried forward the focused morality of Christian tradition that honored individuals as autonomous beings, as independent moral agents, rather than (as pre-Christian traditions typically did) as embedded members of families, clans, and tribes.

A more significant problem for Kant came from his admonition that people abide by rules that they would have apply to everyone. It sounded stern enough, a version of the Golden Rule, but it said little about the content of such rules. It allowed a person to act quite selfishly and ruthlessly so long as he was prepared to have other people act in the same way. As for the rights that emerged out of the recognition of individual duties, their content varied greatly based on the rules of conduct that were crafted. So which rules should prevail?

What became clear in time was that the content of Kant's rules, and what it meant to treat another person as subject rather than a mere object, couldn't be shaped by reason alone, just as Hume had earlier pointed out. The content had to come from somewhere else. Kant believed in God and asserted that individuals should act out of a spirit of goodwill. For Kant these starting points (augmented by speculative logic) seemed adequate. But by the next century, with religious belief on the wane, Kant's religious foundation seemed less sure. The more solid grounding for morals, the only sturdy grounding perhaps, seemed to come from some form of moral sentiments. It came from a deep-seated sense within people about right and wrong, doubtless shaped to varying degrees by genetic inclination, experience, and inherited culture.

Kant's legacy, though, would remain strong, not just in his stress on duties/rights and on the importance of moral rules as such, but in two other important ways: in his insistence that individuals as such were free and fully responsible for their own choices, and in his contention that humans played active roles in interpreting the surrounding world. Human senses and knowledge were limited, Kant asserted, which meant individuals could rightly embrace understandings that went beyond the empirical facts. With knowledge limited we were free to believe, and indeed, he proclaimed, had to believe in order to live morally.

Utility, Parts, and Wholes

Over the course of the nineteenth century, philosophers in various ways largely came to agree that moral principles simply had to arise in some manner, direct or indirect, out of human action, conscious or not. They could not simply be found in nature and could not arise from pure reason alone. Nor were philosophers willing to concede authority to the church or to scriptures or other forms of revelation. A person might simply choose to embrace the church's teachings, as Danish philosopher Søren Kierkegaard would. But it was the individual choice, then, that gave authority to the church. With this stress on individual choice morality increasingly came to seem subjective and personal, a matter of individual opinion based on individual experiences. The liberty rhetoric of the revolutionary era pushed in this direction. So did Kant with his insistence on individual freedom and will to believe. It was an appealing line of reasoning and a venerable one, too, with a heritage reaching back to the ancient Greek Sophists.

Yet, even as they increasingly stressed individual freedom and the power if not duty to choose, philosophers did not lose track of the reality that individuals participated in a social order and had to get along with one another. People formed communities. Somehow the moral order had to sustain the welfare of these communities. Writing in the eighteenth century, Jean-Jacques Rousseau believed that the higher self was one who would (or should) identify with the good of society as a whole. The mature moral being was one whose personal desires and happiness blended with those of the community as such, so that no conflict between the two existed. Writing at the turn of the century Georg W. F. Hegel also retained emphasis on the larger social whole by insisting that the world's parts were all connected, humans included, and that parts could not be understood in isolation. It was essential to consider also their relationships and interactions. The larger issue here—the parts and the whole—soon became central in utilitarian moral thinking, which arose in the first half of the nineteenth century in the writings of Jeremy Bentham and of James and, especially, son John Stuart Mill.

For the new utilitarians, the morality of conduct was best judged not by reference to abstract moral principles or Kantian rules but instead by calculating the effects of an action on human welfare (originally on human pain and pleasure, later on happiness more broadly understood). A moral act was one that brought the greatest net gain in human welfare compared with alternative acts that might be performed. Some versions of utilitarianism would insist that actions comply with rules, and that utilitarian calculations should focus on the comparative consequences of different rules rather than of distinct individual acts. Yet, all versions looked to the consequences of acts to judge their goodness. What was quickly apparent, though, was that this approach seemed plausible only if it took into account the happiness or welfare of everyone; it wasn't sensible for an individual simply to maximize his own happiness and ignore the effects of his conduct on other people. It was clear, too, that one person's happiness often arose in circumstances that diminished the happiness of someone else. How, then, to align the happiness of the individual with the happiness of humankind as a whole? Bentham further complicated his calculations by contending that the happiness of certain nonhuman species ought to be considered as well.

These concerns about the larger social order, about humankind as a whole, tempered the push to expand individual rights and liberties. In some way people had to act as good community members. But where was one to find the moral limits that bound individual freedom, and what made them binding? Bentham's original calculation, based on individual pleasure and pain, seemed to be grounded in empirically testable facts. It was an objective standard, whether a person did or did not experience pain. Bentham merely had to assume as a starting axiom that pleasure was good and pain was bad, nothing more. With the shift, though, from pleasure to happiness and to welfare generally as the operative unit, utilitarian calculations seemed more and more to defer to individual preferences. What made people happy or promoted their welfare? Answers seemed subjective, not objective. This new focus on happiness or welfare also made it harder to compare and sum up the consequences actions had on different people. How did one add up the good and bad consequences of a particular act or proposed rule of conduct when the consequences were based on individual subjective responses and there was no metric to use in measuring and comparing them. And what about actions that made some people happy and others unhappy?

Like Rousseau, John Stuart Mill, the greatest of the utilitarians, hoped that people would progress morally to the point where they aligned their personal preferences with the well-being of the larger community. If that happened, the conflict would disappear. Writing soon thereafter, Karl Marx similarly hoped that the desires of individuals would in time merge with the welfare of the community as a whole as basic human needs were met. Ideally this moral uplift would lead to the disappearance of distinct social classes and even to the end of government (a tool, Marx said, used by the stronger class to exploit the weaker and thus not needed when classes disappeared). Marx, though, was far from firm in making a prediction on this; he merely hoped it would happen. Mill, too, understood that his vision of harmony was based mostly on hope and on his admitted inability to see any other way for individual happiness and group happiness to line up.

At its root, utilitarian thought was chiefly a mechanism that turned decision-making authority on moral issues over to individuals. Their happiness or welfare was what typically counted. Consequentialist moral thought generally only worked when some normative standard existed to judge the consequences. Which consequences were good and which were bad? However up-to-date and mathematical utilitarianism might be, it had no good answer except to leave individuals to decide for themselves and then to sum up their answers. But this was merely a procedural approach to morality. It did not decide *what* was moral; it simply prescribed *who* would get to decide. And this was true even when utilitarianism was put to use—as it was, quickly and often—to criticize institutions and laws and to promote reforms to augment overall happiness. Reformers still had to look to individuals and find out what they wanted.

As for the individuals themselves, utilitarian thought left them free to develop their preferences as they might choose, using various modes of reasoning, religious faith, passions, or mere whimsy. It provided little in the way of guidance except as it implied that they should think of themselves and what made them happy. Plainly it made moral discussion at the social level more constrained. The aim of government and public policy was simply to help individuals as such gain happiness. The good of the whole was merely the sum of the good of the parts in isolation. But where did that leave the idea of a common good, the idea of larger moral or prudential goals that society as a whole might pursue? Where did it leave the age-old idea that morality was a matter of obligations that imposed external constraints on individuals without regard for their wants and wishes?

The Moral Facts of Community Life

Looking back, the patent incompleteness of this moral reasoning as it came together in the nineteenth and twentieth centuries—both the Kantian deontological reasoning and utilitarian moral reasoning—all arose from the rejection of religion and revealed moral knowledge and from what has come to be called the is/ought dichotomy. The basic claim was that the empirically learned facts of the world simply did not offer moral instruction, even when the facts were manipulated using reason. The physical stuff in the world merely existed as such. There was no goodness or badness about it. Accordingly, one could not draw normative conclusions by studying the world. One could not go from the "is" of an existing thing or condition to the "ought" of what should be. One could not go from facts of the world to values; facts and values were categorically different. This dichotomy would be challenged on the ground that human fact-collecting itself was not value-free, which meant that facts as people understood them were necessarily infused with human values. But it was largely agreed that facts, to the extent they could be gathered objectively, were not themselves (that is, taken alone) the source of values. Values required at least some engagement of the human will. Human engagement in turn was largely grounded in human feelings and sentiments, guided and pruned by reason, that is, by the complex mechanisms of the human brain that Freud and others would soon open to the world.

This separation of facts and values hardly meant that facts were irrelevant in making moral judgments. It meant more modestly that the basic values that were used to pass judgment had to come from some other source, even if the values were simple principles (for instance, that humans have moral value, or that human pain is bad). By the twentieth century important moral philosophers were questioning Hume's dichotomy, claiming that the social embeddedness of individuals, their primary existence as social beings, played a key role in shaping moral values. Individuals were not simply autonomous actors. They were parts of larger systems and morality had to do with their roles in the systems. The facts of this embeddedness, they asserted, were themselves infused with values. The is and the ought were not, after all, so distinct.

This stress on the social roles of humans appeared prominently in the work of the leading American philosopher John Dewey. Dewey didn't deny that people were individuals but he insisted on challenging and blurring the presumed line between the individual and society, much as did his slightly older contemporary, pragmatist William James. Individuals were embedded in society, Dewey (and James) claimed. Solidarity—fraternity, others would call it—was as important a value as independence. Dewey also believed, like others before him, that the parts of the social order could not be understood without first grasping the whole and seeing how the parts related to it. Further, the self could be realized only in and through its communal roles. Dewey's thought resonated with important contemporary lines of thinking then understood as conservative, thought that similarly embedded humans in a traditional order (often hierarchical) and that held people accountable for acting responsibly within that order. People fulfilled roles and roles were governed by status rules and expectations.

Dewey's process-oriented thought highlighted the reality that moral thinking could not be detached from thinking about the nature of existence, from the subject of ontology. The drift of Western intellectual thought since the eighteenth century had been in the direction of greater individual autonomy. The American and French revolutions proclaimed it. Kant helped give it a philosophical grounding that Bentham and others built upon, even as they disagreed with many of Kant's claims. Both Kantian and utilitarian thought began with the individual as such and moved outward, exploring how the moral worth of the individual might best be translated into a more encompassing scheme of moral thought. Both approaches stumbled, for the reasons given. But they gained dominance nonetheless because they fit so well with calls for ever-greater liberty (particularly in economic realms). They fit, too, with pushes to expand human rights to cover groups of people long on the lower rungs or social fringes.

By the late twentieth century, more and more observers were emphasizing how individual welfare depended on the types and health of a person's social roles, many following lines of reasoning termed "communitarian." Stark individualism just didn't fit the ontological facts. In moral thought this would translate into a renewed interest in the moral writings of Aristotle, who had similarly portrayed humans chiefly as social beings. Aristotle's ontological understanding led to moral reasoning that emphasized a life of virtue or, more broadly, flourishing or excellence. Morality was not chiefly about abiding by particular binding rules (Kant), or about promoting the sum of individual human happiness (Bentham), but instead about people living virtuously in and among their neighbors and fellow citizens, honorably carrying out their social responsibilities. Late in the century, Aristotle's moral reasoning would gain an articulate proponent in philosopher Alasdair MacIntyre, whose writing directly challenged modern liberalism in its politically varied forms.

This interest in questioning individualism, in renewing the links between morality and ontology, would appear prominently in environmental philosophy. It was in the environmental area where the physical connections among people were most readily apparent. As ecologists had long pointed out, humans were embedded in larger natural systems. They were, as Aldo Leopold had famously put it, not conquerors of the land community but plain members and citizens of it, as dependent on nature for their survival as any other living creature.

These realities about nature's functioning and human dependence have direct moral relevance. They relate also to our ways of dealing with one another given that one person's actions in nature inevitably affect other people. This new ecological ontology has become even more striking as we have learned more about nature's functioning and how human welfare depends upon it, upon nature's "ecosystem services," as the dependence is sometimes termed. Moral living requires respect for these natural processes if only because they make human life possible.

These ecological realities would lead environmental philosopher J. Baird Callicott to issue a direct challenge to the is/ought dichotomy, much as social ecology had earlier led John Dewey to do so. The facts of the natural world, the ecological realities of interdependence, do have direct moral significance, Callicott urged. Humans in fact are parts of natural communities without regard for what they know or prefer. The facts of interconnection directly challenge both Kantian moral thought and standard utilitarian thought. Both presume the basic autonomy of the individual and construct a moral scheme by teasing out the implications of recognizing individual moral worth. Both lines of reasoning show serious deficiencies, however, the moment the abstract individual is reconnected to nature's lifelines. Looking back, Callicott contended, moral thought veered off course after the mid-eighteenth century by overemphasizing the autonomous individual. Particularly when it came to dealings with nature, the better approach was the older one, the one traceable from Plato and Aristotle up to Adam Smith, the one that began ontologically with humans embedded in communities.

Ecological facts, then, did play a direct role in giving rise to values. Yet ecological facts and social facts were not alone enough to flesh out a moral system. It remained necessary to draw upon values from deep-seated sentiments. It was necessary to turn to sentiments about the value of life; about the rightness of taking care of nature for future generations; and about the value of the whole of nature as such along with the special value of its human members. Nature's ways were exceedingly complex. Right ways of living in it necessarily called for serious scientific efforts to learn and make sense of this complexity. In short, reason, facts, and sentiments all had their moral roles to play.

Summing Up

The wide scope of this chapter makes it helpful to organize the concluding points in summary form. They provide foundations for later parts of the inquiry.

The modern age highly values objectivity when it comes to addressing public issues. Objectivity typically means sticking with facts and logical reasoning while pushing subjective feelings and preferences off to the side. This stress on objectivity appears in various forms, including a tendency on environmental issues to turn contentious matters over to science and to expect scientists to explain whether a problem exists. Without question, the need for good scientific facts is quite high. But science is regularly called upon to answer questions that go beyond it, beyond science aptly understood as a skilled effort to gather, test, and interpret facts. When it comes to normative issues, science cannot give answers and should not be asked to do so unless expressly supplied with normative standards to use. This overuse of science extends to scientific methodologies, particularly burdens of proof and scientific standards for accepting evidence.

Looking ahead, the work of finding our place in nature requires that we make sense of science and then put it in its rightful place. Normative issues need to be identified and understood as such, not treated as factual questions. This work includes thinking clearly about which moral and prudential questions should be resolved and acted upon at the community and which are better reserved for individual choice.

In dealings with nature, many policy options must be selected and implemented at the community level. They require people to work in concert.

As for moral thought, our predicament today is fairly plain. We've followed an intellectual journey over the centuries to a place where we have curtailed our capacity to engage and resolve issues at the community level. Our Enlightenment-derived commitment to objectivity leaves us without sufficient tools to exchange moral views and visions and, through rough consensus, to embrace new axioms leading to better public policies. Moral principles, we now realize, are not simply out there waiting to found. Yes, facts are highly relevant in moral thinking and reasoning plays a critical role. But ultimately our moral thinking will be grounded in our sentiments and intuition, which means, far from being pushed away, sentiments and feeling should be given a central place. They should be aired, exchanged, discussed, critiqued, and refined.

Moral orders begin by people formulating and embracing axioms that simply cannot be proven scientifically or logically. Thomas Jefferson's self-evident truths were not supported in that way, nor could they have been. He offered them rhetorically, as moral philosophers had always done and always must do. They gained their legitimacy as axioms, not when Jefferson pronounced them, but later, through the legitimate means of public embrace. People of Jefferson's age and thereafter agreed with his moral assertions; they accepted his self-evident truths and by their choice turned them into axioms.

Along with Jefferson's self-evident truths we might illustrate the nature of moral axioms by considering briefly the claim that human life is morally worthy. This, too, is a rhetorical claim, ungrounded in facts or logic, which has arisen out of Christian teaching and slowly gained acceptance over time. Animal-welfare advocates would have us broaden this now-fundamental moral axiom. Moral value, they claim, should extend beyond humans to other life forms. Members of certain other species are also morally important and we ought to make room for them in moral reasoning. Such a claim, of course, is not really grounded in facts or reason alone (though both are certainly used). Instead it is a moral assertion offered for acceptance as a new collective axiom, much as Jefferson offered his claims. The moral claim can of course be challenged. But it cannot rightly be dismissed on the ground that it is not scientifically valid or not based in facts or logically compelled. If that were the test, the claim that humans have value would similarly fail.

Moral reasoning ultimately arises out of sentiments and feelings mixed together with facts and clear thinking. We might think of this mixture as a heady soup, its elements running together in ways that make it impossible to specify their precise contributions. Enlightenment Era philosophy was right: facts do not *alone* give rise to moral values. But facts play key roles not just in implementation but in the original formulation of the values. Reason, too, must be in at the beginning, if only to clarify sentiments, to put them into sensible form, and to expose them to the realities of the world.

Ultimately values gain legitimacy by social choice, as philosophers have long emphasized. Without a human valuer to create or otherwise recognize it, moral value does not exist in a meaningful sense. That process could attribute value to individual members of other species or to future generations. It could recognize value also in entities—in species as such, biotic communities, and specific landscape features. Value that arises in this way is intrinsic or inherent, value that exists independently of any contribution to human well-being. But this value still rests on human choice, however inspired and encouraged by nature. That reality must be understood, just as it must be known that, because all morals arise from human choice, it is appropriate and essential for humans to be and remain choice-makers.

FOUR

Liberal Fragments

In a world made up of physical things, lacking any immanent moral order or any guiding spirit or Logos, people are essentially free to make sense of the world as they see fit. They can distinguish good from bad and right from wrong. Gazing around the world, they are at liberty to decide which parts or processes will possess value and which will not.

This valuation process, needless to say, has no effect on nature itself—no effect, that is, except insofar as the attribution of value affects how people then interact with nature. A given scheme of human-created values could lead people to use nature in ways that keep it fertile and productive. A different value scheme could allow or encourage land abuse, even as it might bring good results in other ways. It is thus with caution that we recognize our power to decide where value lies in the world. We are free to do so, but nature in a sense is the ultimate judge or at least a demanding and insistent partner. The wise approach, plainly, is to study how nature works, to take into account our needs and wants, somehow to factor in the limits on what we know, and then come up with an overall value scheme. The whole point of such an effort, it needs seeing, only relates to nature indirectly. By attributing value we construct a framework for guiding how people act. When we say that a thing is valuable we say that it deserves better treatment from us than a thing that is not valuable. Value attribution is about guiding human conduct.

Over time many civilizations slid downward because they used nature in ways that disrupted its fertility-sustaining processes or that exhausted key resources. An unwise scale of values can be a root cause of degradation, but there have been other, different ones. Most common, in terms of types of land abuses, have likely been abuses of soils and of the natural processes that build and protect soils. Livestock can quickly overgraze hilly lands leading to massive soil erosion. Most irrigation also degrades soils over the long term (decades sometimes, centuries in other settings). Irrigation is sustainable when it involves diverting silt-laden floodwaters onto farm fields, the type of irrigation long used in the Nile River valley. Irrigation of other types routinely degrades topsoil. As it seeps below the surface irrigation water dissolves salts and carries them upward. Over time, these salts can crystallize on or near the surface, requiring ever-greater amounts of water to flush them away. The process of flushing salts, though, works only for a time. Ultimately the salts win, forming crusts that inhibit plant growth and slow or end crop production. Significant parts of the Mediterranean region and Middle East show scars of both of these types of soil misuse. Today's Mid-East wars cannot be fully understood without giving emphasis to the region's history of costly misuse.

Our valuation of nature, then, needs to take account of what nature entails, how it works, and how we depend on it. As covered in the last chapter, however, facts alone are not enough to give rise to values. Choices are required, concerning the types of lives we want to live and for how long and how we aspire to get along with one another. The relevant issues here are many:

- Do we want to keep land fertile and productive for future generations, without foreclosing options or depriving descendants of chances to experience nature's full richness?
- Do we want simply to protect species that help meet our needs or go further to retain as many species as we can, simply to have them around and because they might benefit us or our descendants in unforeseeable ways?
- Do we want to inhabit lands in ways that provide room for all people to live and thrive or are we instead content to let the few take disproportionate shares?
- As we go about this work, attributing values and setting norms, how should we acknowledge the limits on our ability to learn about nature; how might we, in effect, confess our inevitable ignorance and avoid placing bets with nature we can't afford to lose?
- Going further, if the good life is based on virtue—as philosophers and religious figures have long professed—then how might a sense of virtue and right living enter into all of this; how might it translate into respect for nature and its ways?

In short, what kind of life do we want to live, individually and together, now and in the future, and how might we best translate these visions of good living into norms that distinguish the legitimate use of nature from abuse?

Any sensible scheme of values would surely begin with the reality of our embeddedness in the land community and with our inescapable dependence on it to meet basic needs. It would admit that nature is dynamic and mysterious and our knowledge of it fragmentary. It would pay particular attention to the ways healthy communities give rise to emergent properties that are critical to nature's primary productivity—for instance, to the ability of a species-rich plant community to withstand drought, pests, and disturbance regimes. It would understand that complex life has evolved based as much on cooperation and interdependence as it has on competition and autonomy, particularly in the case of the planet's most successful animal species. And it would give considerable weight to our social natures as humans, to the ways our happiness, self-worth, and flourishing are linked to strong, mutually respectful social links. Ontology, that is, is critical to the value-creating process. To do it well, we need to understand the nature of existence, human and nonhuman, giving emphasis to lines and chains of co-evolved interconnection.

The Moral World of Humans

Western civilization has entailed a long process of creating, refining, discarding, and recreating senses of value, in social and natural realms. The study of value schemes and their social evolution is central to what anthropologists do. They seek to learn how different peoples see and value the world and shape their lives accordingly. They study how modes of living change the surrounding world, leading in some dynamic, dialectic fashion to further changes in human perceptions, values, and modes of living.

A foundational issue for any society is to decide which humans will be deemed morally worthy and to what degree. Nearly all societies have attributed moral value unequally among people. They do so based on differences in age, gender, kinship, prowess, apparent spiritual powers, and other factors, yielding a differentiated social order that is more or less hierarchical. For most of their history humans have lived in tribal groups, commonly (it is now thought) in groups of approximately thirty people. Such groups were typically linked in reasonably peaceful and productive ways with some number of similar groups, with whom they traded for goods and outside marriage partners. Tribal societies, however, tended to demark territorial boundaries that were fairly distinct and to defend them against outsiders. Thus, in the typical tribal moral order, moral status was highest among people within the tribe, existed to a lesser degree in the case of members of associated tribal groups, and might exist hardly at all in members of distant and hostile tribes. Within and between social groups moral status also paid attention to religious ties, ethnic and racial similarities, and other factors rooted in history and genetics.

Only in the modern age—and as one of modernity's great achievements—has it come to be widely said that all humans possess at least a core of moral value. No human still resides on the amoral plane of rocks. Gradations in value, though, remain alive everywhere. For instance, we are prone to value immediate family members far higher than strangers, and perhaps local community members higher than faraway people. But in the common view all humans are morally worthy. It is an extraordinary moral stance, never before known in any widespread way.

Given this novel view the modern age is now characterized by a value scheme that distinguishes rather clearly between morally worthy human life and all other life, between *Homo sapiens* and the planet's other eight or nine million species. Some cultures do venerate particular species; those cases can be cited as exceptions to the generalization. Everywhere many people are strongly attached to their companion animals and the treatment of them can also qualify as exceptions. Finally, laws commonly ban cruelty to certain types of animals, a lengthening list in many jurisdictions. Such laws can reflect value in other animals but may do so only indirectly, when they are motivated by a human-centered concern (such as a fear that cruelty to animals could lead to cruelty to people). The human-nonhuman line, then, is certainly blurred, yet it is nonetheless quite apparent. It supplies one of our central moral principles, guiding how we make sense of the world.

But before we pat ourselves on the back for having attained this lofty moral plane we need to consider how much moral value we actually do recognize in other people. Other people are worthy, to be sure, but what does that mean? Value attributed to people has the same purpose and consequence as value attributed to nature or to anything else (to a rare painting or historic building, for instance). Value is all about guiding human conduct. So how does our recognition of value in all people affect our actual behavior?

An accurate answer is hard to produce. We deem it wrong to kill or physically hurt other people, at least without good cause. To a lesser extent we respect their claims to personal property, particularly to physical items linked to their basic needs or identities. Humans whom we value more highly—family members, neighbors—likely receive more respectful treatment than this. But as for humans as a whole, there are real, practical limits on their worth. Their moral value doesn't routinely lead, for example, to a felt duty to provide food for the starving or shelter for the homeless, though we sometimes do. Nor does it routinely impose limits on high consumption by the wealthy or on how nations and businesses go about grabbing and dominating the planet's best lands. There is something wrong with poisoning a neighbor's drinking water; that seems to violate the ban on direct individual harm. It's not so clear, though, that broadcasting pollution widely clashes with the moral worth of other people. It is not clear that one person's hoarding of valuable resources is morally problematic or that there is anything wrong with an economic system that rewards hard-working people in highly unequal ways.

These comments introduce two topics that are taken up in part 2 of the book. One topic is this blurry moral line between human life and all other life forms. It is assessed in the course of a broader look at how we ought to think about other life forms. The other topic is the expansive matter of social justice and the ways our desires for social justice ought to translate into changes in how we value, use, and share nature.

Here these two topics are put to the side to look instead to other core elements of the modern worldview, principles that structure how we think about good and bad and right and wrong. When we open the door and head out into the world, a world of people and nature, how do we make normative sense of it? What principles shape our public policies and the ways we talk about big issues? In brief, how have we exercised our powers as creators of public value?

The Jumble of Liberty

At the core of the modern scheme of values is the individual human, understood chiefly as an autonomous being possessed of considerable liberty. Particularly in the United States liberty is the preeminent value, so much so that it goes far toward defining what the nation is all about. As such, it provides a logical point of beginning to make sense of our moral universe.

Liberty is commonly understood as an attribute or right that an individual possesses. Most often it is understood in negative terms—as the right to engage in activities free of interference by others. A person's liberty to act is thus linked to limits on what other people can do—linked, paradoxically, to restrictions on the personal liberties of these other people. That the liberty we most value is negative liberty is easy to see. A person who owns a tract of land might be free to build a house on it; he has the negative liberty to do so in that others can't interfere. If he lacks the money for the construction he won't be able to undertake the building. But a lack of money is not commonly thought about as a restraint on liberty. It is something else, less a restraint than a simple and perhaps unfortunate reality.

The lack of money to act as one wants is expressible in terms of liberty. It's about positive liberty, the affirmative capacity to get something done. Liberty comes in positive as well as negative forms just as it comes in collective as well as individual forms. When these forms of liberty are brought together it becomes evident rather quickly that the various types of liberty clash. One person's liberty to take an action that affects another can override the other's liberty to be free of interference. Similarly, a community's liberty to craft rules protecting its natural home can clash with the liberty of an individual to act free of the limitations. Overall, liberty is, if not a zero-sum game, at least something close to it. Abraham Lincoln captured the conflict in a characteristically vivid, slavery-related narrative:

The shepherd drives the wolf from the sheep's throat, for which the sheep thanks the shepherd as a *liberator*, while the wolf denounces him for the same act as the destroyer of liberty.[1]

To proclaim liberty, Lincoln illustrated, was to proclaim nothing meaningful without making clear the types of liberty being exalted and the types of liberty being sacrificed.

A person's liberty to act freely is constrained by the requirement that the actor avoid harming anyone else. This common understanding, no doubt of ancient origin, is often associated with British utilitarian writer John Stuart Mill. His classic work *On Liberty* has stood since the mid-nineteenth century as a foundational apology of liberal individualism. In it Mill stressed the primacy of individual liberty, by which he meant not so much freedom from constraint by government but freedom from constraints coming from the social order. (In Mill's instance, he was irritated by social disapproval of his long, close friendship with a married woman.) One's liberty extended, however, only so far as one's actions caused no harm. This point seemed plain enough, but Mill was a careful thinker. He could see better than others that harm was not self-defining. A do-no-harm norm really meant very little until the concept of harm was explained. If harm included any effect that a person's actions had on other people—even on simple observers—then pretty much every human act caused harm. Mill didn't use examples from nature to illustrate his argument, but he could have: Any significant act in nature triggers changes that alter nature for other people, if only subtly and invisibly. It affects what they can do. If harm includes all effects, then the liberty to act is hardly any liberty at all.

Mill's classic work had philosophic predecessors, particularly John Locke's seventeenth-century *Second Treatise on Government*, an early liberal or proto-liberal classic. Locke's volume is remembered in part for his extended story about the origins of private property, beginning in a hypothetical state of nature. In a state of nature, Locke said, an individual had the right to mix his labor with a physical thing in the world (including a tract of land) and through that labor to create value. Because each person owned his body and his labor (presumably), the value created by that labor belonged by natural right to the laborer. And when the value of the ultimate thing was due entirely to that labor—or perhaps due 99 percent to the labor—then by natural right the laborer ought to own the thing itself; it became his private property. There was no injustice to this arrangement, Locke claimed, because any other person who wanted to go out and labor in the same way could do so. One person's actions didn't limit what others could do. This was true, however, only on one quite important condition, a condition that Locke emphasized in his famous "proviso." A person could claim ownership of a thing, Locke stated, provided that there was ample opportunity for others to do the same, only if the thing was essentially unlimited in supply and thus lacked value in its unimproved state. More generally, private property in Locke's view could arise only so long as one person's claim of ownership caused no harm to other people. This was Locke's do-no-harm limit on liberty, and Mill built upon it.

As commentators on Locke have long noted, Locke's narrative about land ownership made no real sense. Land and other resources were scarce everywhere. One person's control over a piece of land always limited options for others. Claims of private property—at least in land and natural resources—always caused harm. Only if one stuck to

Locke's fanciful world of abundance could his story legitimate private property rights, at least in anything other than the trivial acorn that drops to the ground surrounded by countless other acorns. Once real-world scarcity is admitted private ownership raises serious issues.

As so often is the case, however, Locke's friendly readers were not inclined to get picky about the facts. Locke's *Treatise* supplied a grounding for private ownership that didn't involve an initial land grant from the king (the whole point of his writing). That work done, Locke had satisfied his core constituency (Parliament, mostly) and could bask in their admiration. Locke, though, had highlighted the importance of the do-no-harm principle and had admitted that it was a morally necessary limit. Locke could see no way to justify harm-causing conduct based on the natural order alone, not without positing a social order and some form of collective decision-making (which for various reasons he wanted to avoid).

Sensitive to this issue, John Stuart Mill was quite different from ordinary readers of Locke's famous treatise. As Mill saw things, liberty and do-no-harm allowed people to have complete freedom of thought. They could think whatever they wanted without causing problems for anyone else. At the time, this freedom was not at all taken for granted. Many public offices and other perquisites (university posts, for instance) were limited based on religious belief. Yet Mill could see that even translating thought into words, turning ideas into speech heard or read by others, could bring about harm. This meant that the liberty he justified and defended did not necessarily extend even to freedom of speech, much less to any freedom of conduct. Read closely, in fact, Mill's tract was hardly a ringing endorsement for any broad sense of liberty. As for what qualified as harm—the key issue, Mill could see— the only logical answer was to view harm as a social construct. It was society as a whole that gave the term content, thus setting the bounds on individual liberty.

The quandary that Mill faced is still very much with us. Individual liberty is widely proclaimed, usually with mention made of the obligation to avoid harming others (and typically also with a note that others should enjoy equal liberty). But without a definition of harm this reasoning is largely empty of content, as Mill could clearly see. As for the definition, harm can't be defined simply by drawing upon the multi-sided notion of liberty itself. The meaning of harm has to be based on other normative values. In practice, its meaning is set by society through its tolerance or intolerance toward particular types of behavior. That tolerance rightly evolves over time along with social values and understandings.

Harm and Nature

This vagueness of the do-no-harm rule—or more positively the social responsiveness of it—is central to the ways we think about nature and how to use it rightly. It also highlights one of the ways contemporary culture could easily shift to increase protections for nature, without coming up with a new moral frame. When harm is defined so as to include ecological degradation, then the do-no-harm limit stands ready to offer protection.

As commonly understood "harm" means harm to other people—or even more narrowly, to people who can be personally identified and whose harm is visible. In some imprecise way, harm is also limited by a requirement of materiality; harm exists only if it is in some sense substantial, not just trivial. Beyond that, there is a tendency to look at the harm-causing action and ask whether it involves a type of activity that people typically undertake. Is it an activity that an ordinary person would view as legitimate? If it is, then the resulting consequences are usually not deemed harmful. These limits on what qualifies as harm are central to much of land-use law, particularly the law of private nuisance. Harms that are not substantial enough, harms that result from ordinary activity, are simply tolerated as inevitable costs of living in a congested world.

As thus defined, then, harm is typically limited to harm to another person. Nature is harmed in a legal or moral sense only when the affected part of nature is someone's property. The true harm is to the property owner, not to nature directly. As for how material the harm is, that too is typically judged by the effect on the property owner, particularly its economic effect. When harm like this takes place, the proper remedy is to compensate the owner for the economic loss. The remedy is not to the land itself, nor is the harm really measured in natural terms (for instance, in changes to the composition and populations of the resident plant and animal species). Understood this way harm doesn't pay direct attention to or value the interconnections in and among natural systems. It doesn't pay attention, directly at least, to future generations, to other life forms as such, or to disruptions of the emergent properties of natural systems. Perhaps most starkly, it largely ignores the effects a property owner's activities have on land that the actor owns. Land-use rules change this overall picture somewhat, but the basic premise still is that the ownership of land includes the right to consume or degrade it. Harm means harm to neighbors.

These comments on harm highlight ways that today's moral thought discounts nature and sanctions misuses of it. Land abuse by a private owner is shielded by property law, which incorporates a narrow definition of harm and is grounded in individual liberty. These comments also make apparent—particularly the vagueness of the idea of harm—how fragmented the idea of liberty is as a moral principle to guide conduct. Without a definition of harm, we can't know when one person's positive liberty to act should take precedence over another person's negative liberty to be free of interference. We can't know when the collective liberty of people to get together and organize their joint affairs should take precedence over the desire of an individual to escape the group's decision. Liberty alone provides no guidance, and no amount of flag-waving or rooftop shouting about it can resolve its deep ambiguity.

At bottom, liberty is best understood as a way of talking about *who* gets to make a moral decision. Who gets to choose how to use a particular tract of land? Who gets to decide whether a wild wolf will wander freely or whether a stream will continue to flow without human disruption? As we go about choosing among competing types of liberty, as we go about defining the term "harm," we necessarily answer these questions and allocate power. Is the private owner of the land the one who should decide? Is it the local community? Is it the national or even international community?

Liberty is a useful enough concept, particularly when defined clearly. There are good reasons why it is so much honored. But it is not a moral principle that, standing alone, can steer our way toward better ways of living in nature or with one another. Meanwhile, the shout to protect liberty goes on, most often as a call for still-looser definitions of harm (contentions, for instance, that filling a wetland or high-chemical farming is not harmful). To be sure, a landowner given considerable latitude might exercise her expansive liberty in ways that promote the land's health. But the call for liberty as such largely pushes in the opposite direction. It is a call for individual freedom to alter land as one sees fit. It is a shield to ward off and silence outside critics. As a moral frame, liberty exalts the individual human in isolation, not the human who really exists, not the human embedded in social and natural systems.

The Fragment of Equality

Less needs to be said about the core moral principle that is so often coupled with liberty, and that is the principle of equality. Together liberty and equality go far toward shaping the modern moral order, once again most prominently in the United States, which sees itself as the land of liberty and equal opportunity. When probed, equality, like liberty, has only modest moral content to it, valuable though it is. In the vital, age-old task of trying to figure out how we ought to live on land, equality offers precious little help.

In a formal sense equality means that like cases should be treated alike, particularly in the law. Lines should not be drawn among people or cases leading to different treatment on a basis that is arbitrary. This principle is not an intellectually empty one, as critics of it have sometimes said. Without the principle of equality it would presumably be fine to draw arbitrary distinctions, to give different treatment to cases that are essentially identical. Hierarchies in which people are fixed in place, with vastly different treatment, might well be acceptable. Beyond that, though, equality has little content to it unless linked to something else, unless it builds on some understanding of human moral worth.

Equality's central shortcoming—a big one—is that it has no formal means for deciding whether particular cases are or are not alike. It has no means for evaluating whether a distinction drawn among cases is arbitrary or, instead, based on a morally significant factor. Similarly it can't say whether a rule or condition that, on its face, draws no distinctions nonetheless violates the ban on inequality because it leads to different consequences for people in varied circumstances. For instance, are people treated equally when everyone who enters the courthouse must climb up ten steps to do so? In one view the courthouse steps provide equal access to all, no exceptions. In another view, the steps give unequal treatment to those who can't climb them. So which is it? The principle of equality itself can't give an answer and that is its grave weakness. An answer must come from another moral source, from a scale of values having to do with an individual's ability to participate fully in society and with senses of belonging and self-worth. From its emergence in the Middle Ages (pushed by church writers on natural law), equality was in fact linked to a sense of moral worth attributed by God. But equality didn't define or give content to that moral worth; it merely extended the moral worth to all people. The moral worth itself was a prior and independent moral stance, historically based (in the West) on Judeo-Christian scriptures.

Aside from its limits, equality can generate problems or confusion in our dealings with nature. They crop up when owners or users of nature challenge rules limiting what they can do, rules attacked on the ground that they treat people unequally. Landowner A asserts the right to develop land by citing landowner B's ability to do so. The law should treat them equally, she asserts. If in fact different rules are put in place specifically based on the identities of the owners, and assuming their identities are not really relevant, then the law might well operate unequally. More likely, though, is that the law doesn't treat people as such differently. It treats land parcels differently. And A's land might well differ from B's land in a way that justifies disparate treatment. This observation is evident enough, and confusion should hardly prevail. But in the property-rights arena in particular categorical thinking is never far from the surface. A person who seeks to develop today can be outraged if a neighbor owning quite similar land was allowed to develop ten years earlier. Shouldn't property rights be timeless? Projects ten years apart in time, however, might be appreciably different in ways that law can rightly consider. Social values may have shifted and social definitions of harm along with them. The need for treating like cases alike remains strong. But which cases are really alike?

The moral prominence of equality is connected to—and in fact, almost a subset of—the principle of the rule of law, the idea that public affairs should be guided by law rather than by arbitrary exercises of power. It is a powerful and important ideal, even as it evidently leaves open questions about the content of sound laws. We want a government of laws, not of autocrats, but what kind of laws? Without more moral guidance we cannot know. This was the grave uncertainty embedded in Immanuel Kant's philosophic call for people to live in accordance with the moral principles that they would apply to other people. Different people can propose much different rules, all capable of being applied uniformly.

In short, equality like liberty is a fragment of a moral vision, useful so far as it goes but no further. As for liberty and equality in combination, they fit together easily enough. John Stuart Mill's limit on liberty—his do-no-harm rule—was set forth as a rule of equal application. But even together the two principles say almost nothing about how we ought to interact with nature, just as they provide only fragmentary guidance for social interactions. They provide little help in deciding whether particular changes to nature are or are not abusive. Much more is needed.

The Contingency of Rights

These comments on liberty and equality introduce the topic of human rights generally, and how they fit in with the labor to live sensibly on land. Human rights, as noted, are social creations based on choices in some way made by and among the people who recognize the rights. The content of rights is rather directly linked to the reasons why they were created and the work they were intended to do. When John Locke came up with a natural-rights justification for private property his aim was to counter the claims, made by England's Stuart kings, that all property arose by grant from the crown and was thus derivative of royal power. Not so, Locke said. Property could have arisen, hypothetically at least, by people mixing labor with land under carefully defined conditions.

The rights held dear in the United States were largely clarified and embraced during the republic's early years, and they arose out of the revolutionary conflict with Britain. Like Locke's property rights, America's founding rights were crafted for particular reasons. The evil of the hour was a distant, intermeddling British state that sought to impose its will on the colonies. Various revolutionary forces were at work, many of them economic. Many leading citizens wanted simply to shift political power from England to the colonial capitals so that colonists collectively could rule themselves. Many others, in lower and aspiring social groups, nurtured a desire to get free of a broader array of legal and social constraints.

The immediate need in this revolutionary context was for an understanding of human rights that somehow confined the powers of the distant government. The rights thus chosen, appropriately, were defined in those terms, limits on the powers of government to control what people did. They were not, in contrast, rights that colonists could assert against one another (servants against masters, wives against husbands, or fishers against mill-owners throwing up dams on streams); indeed, one strand of Revolutionary-era thought was distinctly communitarian. In a predominantly agrarian world people saw little need for government above the local level. Assertions of political rights against centralized government thus fit the needs of the hour and came at little cost.

The human-rights moral order crafted in the Revolutionary era worked well enough for a century and more. The federal government was relatively insignificant except when fighting wars. State and local governments promoted expansion and economic development in ways that fit the temper of the age. Rights-thinking helped fuel efforts to expand public education, to reform mental institutions and prisons, and vest property rights in married women. In time, though, rights formulated to deal with a distant government proved ill-designed to address the rising problems of powerful economic forces and concentrated wealth. When confronted by private power and feelings of being pushed around, ordinary people found little protection in political rights based on eighteen-century conditions. Indeed, Revolutionary-era rights mostly seemed to help the other side. Citizens could deal with the rising economic forces only by working collectively, drawing upon the powers of government. Yet their human rights as put into law largely took the form of limits on governmental power.

The situation might have been different—the cultural context surely would have been different—had America's notion of rights been forged at a different time and under different circumstance, if the nation had, for instance, started not with Jefferson's "life, liberty, and the pursuit of happiness" but the alternative phrasing used in the founding of Canada, also taken as self-evident: "peace, order and good government." By the twentieth century, nations crafting their own lists of rights would go beyond the nineteenth-century Canadian charter in stressing positive rights and elevating government as a beneficial tool. They added rights to housing, education, and clean water, rights far different in form and temperament from the fewer, dated rights possessed by Americans.

Particularly in their American form but also to varying degrees in other forms, rights tend to center on the individual human as such, considered as an autonomous being. It is rhetoric that is not merely human-centered, paying no attention to other life forms, and not merely present-centered—only people living today count—but based on a vision of human existence that discounts social roles and the larger community of life. Rights uphold the human parts as parts. Their incompleteness is partially explained by the fact they were intended when embraced only to supplement the then-prevailing moral order, largely drawn from Protestant Christianity. Religion-based morality provided the frame for public life as well as private life, supplemented by political liberties. As for nature, it was still viewed in North America as essentially a limitless cornucopia. It was a continent of vast natural wealth with little apparent need for any limits on using it.

Rights in America are incorporated into constitutions and draw their legal power from them. Americans, however, often view their rights as transcending federal and state constitutions, as somehow originating apart from them and merely confirmed by them. That view of rights, though, is hard to justify except as a type of political creation myth given that rights are highly dependent on historical and cultural circumstances and take effect only by public acceptance. Even as he spoke in favor of liberty John Stuart Mill made clear that human rights were entirely social creations, not objective ideals embedded in the universe from its creations or perhaps miraculously arising (by divine act?) when humans first walked the planet. Moreover, rights-claims were legitimate, Mill stressed, only when and insofar as their recognition generated overall benefits for the common good. Individual rights were derivative of public welfare, he insisted, and they should be crafted to help support that welfare.

America's pragmatists at the turn of the twentieth century—jurist Oliver Wendell Holmes Jr. prominent among them—agreed with Mill on these points. Societies as such created human rights, and they did so when the embrace of them promoted the common good. Holmes was particularly inclined to judge rights-claims pragmatically. What were the consequences of recognizing a particular claim of right, he asked? Would its embrace lead in practice to good social consequences? Human rights, in short, were the product of public governance, they were outputs, not points of beginning. Once the common good was identified, rights could be defined in ways that helped promote that good, particularly but not only the parts fostered by responsible individualism.

From today's perspective and given looming problems, there is value in seeing more clearly that rights arise out of collective social choice and that they should continue without change only so long as they make good social sense. Rights are crafted to help people deal with particular problems and to flourish under circumstances prevailing when written. In this light it is unsurprising that South Africa's constitution, written as the country moved beyond its apartheid era, contains rights far different from those used in the United States. They take account of South Africa's much different circumstances. They also pay attention to the realities of the modern era, including resource scarcity and ecological degradation.

The final chapter takes up, as one reform strategy, the possibility of revising America's core list of rights to draw in concerns about ecological degradation. Several individual states have done this. They recognize in their state constitutions individual rights to clean or healthy environments already (Illinois, for instance, in Article XI of its Constitution of 1970). Though sweeping in scope and sober in tone these constitutional provisions are little known to people and most courts have treated them as largely meaningless. To have real effect environmental rights would need to be far different in form and substance from the rights that Americans know and cherish. Indeed, the whole idea of rights would require recasting.

American-style rights were intended mostly to check government and to create spheres of individual autonomy free of state invasion. They remain valuable for the particular work they do. But they help hardly at all in the oldest task of living well on land. Indeed, when they come into play they are by and large roadblocks, used to shield continued land degradation or to demand that landowners and polluters be paid to change their harmful ways. When drafted in the eighteenth century these rights helped people to live well and prosper. They can remain morally legitimate—we should continue to venerate them—only insofar as they continue to do so.

Economics and Objectivity

This chapter has taken a look at the moral frames and vocabularies that predominate in discussions about public policy, the values used in public discourse rather than the richer language used to talk about how people ought to behave privately. That latter language of private virtue is laced with concerns about manners and mores, with living honestly and treating other people with respect, and often with personal piety. When it carries over into the public realm this private morality mostly adds further emphasis to the dignity of the individual. It supports a call for society to assist the downtrodden along with a call for greater virtue from the slovenly. In all these roles, however, private morality remains mostly centered on individuals and families as such, less often on any common good that transcends individual welfare. It is a useful supplement to a well-considered public morality, to a moral order that also supports social and natural communities, but it doesn't supply the moral materials to construct such a public order.

Aside from the public talk of liberty, equality, and rights—and putting to one side issues of national defense and keeping the peace—the chief way of talking about the common good is now the economic one. Collectively we gain when the economy improves, so it is said.

Indeed, the push to promote the economy has become something like an all-encompassing goal for most of what government does. What can be said about this big-picture goal as a system or frame of values for distinguishing good from bad if not right from wrong? What sort of moral scheme does it presume or put in place? And how does it fit together with the moral emphases on liberty, equality, and human rights generally?

We can seek answers to these questions by looking at how economists talk about overall welfare. The central measure of welfare, certainly in public economic talk, has to do with the overall size of the market economy. A larger economy is normatively better than a smaller one; that's the message. Measures of the economy typically pay attention to the amount of economic activity taking place, to the value of goods and services being exchanged. In such measures, no judgment is passed on the goodness or badness of the goods and services themselves. The measures are also constrained in that they ignore nonmarket activities and effects, a deficiency that more specialized economic measures partially overcome. High on the list of important factors not considered is the effect humans have on nature. When forests are cut and the trees sold, the market price of the cut trees is included in the annual measure of economic activity. The fact that the trees are no longer growing, that they are no longer part of the living forest, is not relevant to the calculation. The economic measure includes the full value of the timber harvested, without reduction for the loss of the standing trees. Pollution dumped in a river similarly does not enter into the economic measure. Money spent cleaning it up does.

Grounding these calculations is the assumption that people who sell their time and property and spend money are doing so to satisfy their preferences. The market value of what they give up in money and time is presumed to be a sensible measure of the value of the preference-satisfying benefits they receive. It is, by and large, all about the satisfaction of individual preferences. When it comes to valuing goods that are not bought and sold in the market, efforts are made to guess how much people would pay for the good if they did have to purchase it in the market (or what they would accept in money to part with the good if they already owned it). This shadow pricing attempts to measure, in nonmarket settings, levels of preference satisfaction so that these nonmarket goods can be included in calculations of overall welfare. When it comes to alternative ways to satisfy preferences, the better approach is the one that entails lower cost, the one that is more efficient.

Efficiency thus becomes an allied normative goal. More efficient is normatively better than less efficient.

What is useful to note about this approach is that it is, in form, essentially a procedural one and appears objective, even scientific. Economists are not deciding what is good and bad for individuals. Individuals decide for themselves. Individuals display their normative choices by spending money, by relinquishing labor or property, and by responding to surveys with hypothetical questions about how much they would pay to get some nonmarket good. Individuals as such make their own choices based on normative values that they have personally selected. The economist simply gathers data about these preferences and aggregates them. It is this seemingly neutral role (and other work like it) that gives economics its claim to being a science and explains why the field's Nobel Memorial Prize is given for economic science.

It is intriguing that the economics profession overall would want to be viewed as a science. The impulse makes sense given the modern age's cult of objectivity and high elevation of science. Science, as we have seen, is about generating and testing facts. Economics does this when it tabulates individual preferences without passing normative judgment on them. But if it were purely a science, economists would merely lay down their facts without comment or recommendation, admitting that they had no tools of their own to pass judgment on the aggregated preferences. They would step back from their prominent roles as policy advisors. Just as an ecologist as such could not tell us whether a tallgrass prairie is better than a soybean field, so too an economic scientist could not tell us whether one type of living or set of economic conditions is better than another.

This seemingly objective approach to measuring overall welfare is both revealing and troubling. It is revealing because it provides additional evidence of our contemporary cult of objectivity and our inclination to avoid engaging with normative questions directly. With economics understood as a science we can talk about economic growth and how to promote it and still remain in the realm of objectivity. We can promote growth while still leaving normative questions to individuals as they go about working, buying, and selling.

This objectivity, though, carries a high price tag, particularly when it comes to nature, future generations, and many aspects of social justice. If we won't talk publicly about basic values and ideals, then how are we to bend our overall trajectory to a more responsible direction? Further, the use of welfare calculations that pay no attention to normative values suggests that all choices are equally good. In welfare calculations there is no such thing as better and worse or right and wrong; there is simply the matter of more rather than less. When it comes to living in nature, the reality is far different.

Mismatches, Consumers, and Citizens

In fact, though, this popularly familiar economic view of the world is far from normatively neutral. Economists may not expressly pass judgment on the ways that people spend their money or sell their labors; they merely count and measure transactions. But the framework they employ reflects and accentuates a quite distinct normative view.

For starters, the view of the world reflected in economic measures is overtly human-centered and present-centered. Only living humans count because only they seek to satisfy preferences through market transactions. It is also a world in which people are understood as individuals, not as members of social and natural communities. In this familiar ontological view communities as such disappear—their condition and welfare is ignored—and whether individuals act responsibly within communities is a matter of personal choice.

As for the individual preferences, they are, of course, significantly shaped by the market itself and modern advertising, which accentuates selfishness and base impulses. This manipulation of preferences is hardly a practice for which economists are responsible. But economic calculations give it little heed and have trouble, for instance, dealing with buyers who quickly regret what they have done (though some try to do so). In the case of many people, their preferences are also shaped simply by the circumstances in which they live and have lived, by their senses of what is possible for them and the typically limited opportunities that seem open. What people want is constrained (often greatly) by the limited options available and by what they think they can get; it is not about preferences detached from culture and the market. Again, economics is hardly responsible for this distortion but its publicly distributed measures ignore key social realities.

More significantly, when people spend money in the market they mostly do so to gain goods and services that they can enjoy personally or with family. If they are doing the paying, they want the benefits. People are far less inclined to purchase goods and services when the benefits are also enjoyed by other people. When this does happen, when an individual pays the cost but others get most or nearly all the benefits, costs and benefits are out of alignment.

Misalignment like this is of critical importance when it comes to humans and nature because many of the ecological goods that people might prefer—ones perhaps high among their individual wants—are ones that they cannot acquire without benefits also going to other people. A person might be willing to spend time and money to clean up a river, but the benefits of doing so would spread widely among all people who use or enjoy the river. A person might be willing to drive a car with far lower pollution emissions but the improved air quality is not a benefit that the individual can capture.

This mismatch of costs and benefits helps explain the important but little discussed dichotomy that exists between the individual acting as consumer and the same individual wearing his hat as citizen. As a consumer (a market participant) the individual decides how to spend time and money. As a citizen, the individual—ideally at least—decides what public policies to favor, policies that would apply not just to the individual but to others as well. These roles are quite different and the choices people make in these different roles can vary widely.

Acting as an individual—acting rationally, as economists would say—a concerned individual might forgo cleaning up a section of river or purchasing a low-emission vehicle simply because the resulting benefits that the individual can capture (in the form of a cleaner river or air) are trivial. The actions may generate important benefits, but the bulk of the benefits go to other people, the many others who benefit from the cleaner river or the lower emissions. For the individual as consumer, the costs exceed the captured benefits and the action is irrational. If given a choice as a citizen, however, the same individual might support the use of tax money to clean up the river and support a law demanding that automakers reduce air emissions in their vehicles. The position is rational because, in such cases, other citizens are also participating. The individual supporting systemic change now enjoys not just the benefits of her own behavior but the benefits of the similar behavior of all other citizens. The costs are spread among all citizens, and the benefits mostly enjoyed by them. Costs and benefits come into alignment. As consumer the individual does nothing to help the environment; as citizen she is willing to commit.

This consumer-citizen dichotomy is hardly unknown to economics but it receives too little emphasis. It does not, as it should, call into question measures of economic welfare that look only to what people prefer when acting as consumers and market participants—acting, that is, when many collective-action options are not available to them.

In popular, political-conservative forms of economics this dichotomy is largely denied. A common conservative claim is that people express their true preferences only when spending their own money as individuals. What they might want as voters, or what they might say on shadow pricing surveys, is not reliable because then they are largely spending other people's money. This claim is simply wrong, and for the elementary economic reason just given. Countless citizens, for example, regularly vote in favor of higher taxes for public schools even though they do not and will not attend the schools and have no family members who will do so. They support the school tax as citizens because it is good for the community as such and their fates are tied to the welfare of the community. Many will do so—vote to increase their taxes—even though they would not voluntarily contribute an equal amount of money to the school system as a charitable gift. In the case of such a gift, the money would likely bring only a tiny improvement in local schools, a benefit that would go to the entire community, not just the donor. When instead the individual as citizen votes to raise taxes, others will also be paying the taxes and the benefit for the individual is vastly greater.

Economics in its welfare calculations includes public expenditures. Aggregate totals include money spent by taxpayers collectively to purchase goods and services. The central measuring methodology, however, remains strongly centered on what people want as individuals. It does not ask about what would be good for people collectively. It does not measure the overall condition of communities as such using an all-things-considered normative test. When a factory shuts its doors and the surrounding area slips into decline, with housing values dropping and public infrastructure going untended, economic calculators look the other way. They look away, too, when mining companies spill their acids into waterways and destroy aquatic life. These costs do not relate to market transactions by individuals. They are thus ignored, necessarily implying, in the objective opinion of economics, that the consequences are irrelevant to the economy and thus to overall welfare.

A final limitation of economic thought in its popular, public forms is that it provides no good means or vocabulary for talking about the common good as such. It provides no forum for asking about the type of society we would like and how we might promote community health. It is, by and large, only about individual market participants getting as much as they can, goods and services that they can enjoy individually without sharing. Many beneficial goods, however, are ones that are necessarily shared with others. And many can be acquired only when people act collectively, sometimes through voluntary organizations, quite often only through government.

With its facial neutrality—masking a significant normative slant—popular economics is reduced to talking about efficiency as if efficiency were somehow an overall value or goal that society might pursue. But efficiency is not a meaningful goal, not really an end. It is instead a trait or characteristic of means used to achieve an end, often a desirable trait, sometimes not. The call for efficiency routinely confuses means and ends making it harder to think clearly about both.

Moral Bits and Pieces

When the normative ideas and frames covered in this chapter are brought together they seem at first glance to provide a reasonable set of tools to frame public talk about values and goals. We have the recognition of humans as special moral beings, plainly a solid building block. We have the emphasis on individual liberty and equality with the admonition to avoid harming one another, along with the various other human rights that individuals possess. And we have the desirability of satisfying individual preferences in the most economically efficient ways. Yet when the pieces are looked at critically they are less formidable then they might seem; all are fragmentary and make sense only when fit into a larger moral order. When put together they display great gaps, particularly when it comes to our roles and needs as members of the land community.

The uniqueness of human life is a value that promotes human dignity but it does little good when it comes to our dealings with nature. Indeed, the moral emphasis on humans suggests that other life forms have little or no value.

Liberty considered in all its forms is, as seen, nearly a zero-sum game in that more liberty of one form usually comes with less liberty of another form. These days we have trouble thinking clearly about these liberty tradeoffs and don't have a good way to make them. By stressing individual negative liberty—the liberty of the criminal class, it has been termed—we give preference to the form of liberty that is most likely to lead to ecological degradation, if only because legally permitted degradation is a common byproduct of making money.

Equality is an important but fragmentary ideal, one that encourages people to ignore differences. It too doesn't help much when thinking about nature. It can actually hamper clear thought when it shifts attention away from nature's great variations—no two land parcels are ever equal—and focuses attention instead on people (property owners often) and on their demands to be treated equally as people.

Finally, the call in the economic realm to enhance preference satisfaction is helpful only when people prefer to take good care of nature. They are most likely to express that preference, though, as citizens rather than market participants. Popular economic thought gives primacy to the latter role, particularly when mixed (as it often is) with antigovernment sentiment and with overstated claims for the market's greater efficiency. Aggregate economic measures are not particularly sound as indicators of how we are doing collectively, especially vis-à-vis nature. We need to be far more aware of their limits.

Our difficulty in making sense of the oldest task in human history has much to do with the dominance these days of this mix of values and modes of thinking, the limits on them and vast gaps among them. We are all about rights and liberty and equality; about raising up the plight of society's outcasts, about promoting economic growth. This is the language of our day. Its strengths are our strengths. It deficiencies are ours as well.

FIVE

An Ecological Foundation

During the first half of the nineteenth century, courts in the United States tended to refer frequently in their rulings to foundational legal principles. The principles were not binding precedent as such. Instead they were useful, shorthand ways to highlight core elements that structured the law and guided its application. Two principles, both phrased in Latin, were commonly employed in land-use disputes and when landowners claimed that land-use regulations violated their rights. One principle, the preeminent one according to legal historian William Novak, was *solus populi supreme lex est*: the health or welfare of the people is the supreme law. The second was more focused, *sic utere tuo ut alienum non lædas*: so use your own (property) as not to injure another's (property). The latter principle was, of course, a variant on the idea that figured prominently in the liberal thought of John Locke and John Stuart Mill: an individual should enjoy liberty so long as no harm resulted. It was a vague and thus flexible principle, one that could be applied according to local circumstances and prevailing sentiment.

Courts used this language when responding to claims that landowner rights were being impaired. Sometimes property owners won such cases, often they did not. Property rights were highly regarded at the time; that was not in doubt. But individual rights existed in a larger legal and moral framework. As perhaps the era's leading state court justice explained, individual rights needed to be weighed along with the similar rights of the local community. Justice Lemuel Shaw put it this way:

We think it is a settled principle, growing out of the nature of well-ordered civil society, that every holder of property . . . holds it under the implied liability that his use of it may be so regulated, that it should not be injurious to the equal enjoyment of others having an equal right to the enjoyment of their property, nor injurious to the rights of the community. All property in this commonwealth . . . is derived directly or indirectly from the government, and held subject to those general regulations, which are necessary to the common good and general welfare.[1]

Shaw's framework was employed earlier in a ruling by the US Supreme Court by Chief Justice Roger Taney. The dispute involved the alleged sanctity of a charter that granted a monopoly to a company operating a toll bridge across the Charles River near Boston. The importance of private property was not in doubt, but again such rights were subordinate to the larger public welfare:

While the rights of private property are sacredly guarded, we must not forget that the community also have rights, and that the happiness and well being of every citizen depends on their faithful preservation.[2]

Over the course of the nineteenth century language of this type appeared less and less frequently in court opinions. The *salus populi* principle in particular fell rapidly from favor and rarely appeared in rulings by the century's end. Courts similarly found fewer occasions to refer to rights possessed by communities as such. Indeed, as the twentieth century began rights were things that almost by definition attached only to individuals or other legal persons. The *sic utere tuo* principle continued to enjoy favor but its application shifted in response to the nation's industrialization and urbanization. Literally applied, the do-no-harm principle could bring a halt to intensive activities that generated economic benefits but did so by imposing costs on surrounding owners. Do-no-harm tended to favor the first landowner in a region, often a farmer, given that the land-use conflict only arose when a later land user arrived and began operations. The later actor was the source of the conflict, courts originally said.

Over the decades courts paid less attention to *sic utere tuo* as a general principle. Instead they focused on the specific legal rules being applied, typically causes of action in public and private nuisance. Under them, an individual owner or a community harmed by an intensive new land use could halt the damaging use only if it amounted to a legal nuisance. With a bit of tweaking courts revised nuisance law so that it only protected against substantial harms, not harms that were modest. And increasingly they expressly reminded readers that defendant land users also had rights themselves, rights, they now said, to use their lands freely so long as they acted in ordinary and reasonable ways. If they did so, if their activities were similar to the land uses of other industrial-age landowners, then the owners were not engaged in nuisances. The resulting harms were therefore permissible and the neighbors harmed simply had to suffer the consequences. Do-no-harm, that is, became more of a specific rule of law and it came to mean, not avoid *all* harm or even *all appreciable* harm, but instead avoid causing substantial harm under circumstances that were deemed unreasonable under the era's pro-growth sentiment.

The trend taking place, we can easily see, was a trend to allow more intensive land uses, industrial and urban mostly. Necessarily that meant lessened legal protections for the quiet users of land who were being disrupted by their new noisy or polluting neighbors. It also involved a greater focus on the individual rather than on the community. Public nuisance law still existed and it still served to protect public health and safety. But it too increasingly was a specific body of law that existed side by side with other bodies of law. Except in specific settings private citizens faced real hurdles in challenging a land use as a public, rather than private, nuisance. It was the job of civil authorities to bring such legal actions, courts increasingly said, and if civic authorities preferred to side with the new industrial landowners (as they often did), then no public nuisance action would be brought and the harm-causing activity could continue.

What was fading in this pro-industrial shift in property law was the broader venerable notion that the public welfare was not just important but supreme. Disappearing too was the idea that communities as such had legal rights, or even that they existed in the law except as incorporated entities. In moral terms, the individual was ascending in primacy, in law just as it was in the moral philosophy of Kantian deontological thinkers and utilitarians. Rights-based thinking would gain further ground in American law over the ensuing decades. Individuals held rights as autonomous beings, it was stressed more often, rights that were detached from social settings. In the eyes of the law individuals were not embedded, not overtly at least, in a shared moral order that valued and upheld the welfare of communities as such.

Individualism Ascendant

In its various guises this individualistic perspective, in land-use law and public morality, is essentially where we find ourselves today. The moral order is made up of rights-bearing individual humans. They carry their rights with them wherever they go. Where they live is of no moral consequence, nor is their communal embeddedness. Social communities as such have faded in moral and legal importance. Natural communities as such never carried much weight (though valuable parts of them certainly have). The moral focus is on the present, largely ignoring future generations. Other life forms gain value only as private property, and it is value chiefly measured, in the law at least, by market processes. Government's main role is to keep public order so that individuals can go about exercising their rights as they see fit.

In this present-day, individual-centered cultural view, the various goods that people might desire are quite often thought of as marketplace commodities, things that people might buy and consume as individuals and families. Those who want a particular good or service can turn to the market for it. A good school for children is something parents with money can buy. A safe neighborhood, one with parks and sidewalks and public services, is also for sale to those who can afford to move into it. For those who want such goods the dominant message is abundantly clear: earn money and then go into the market to the buy the goods.

How this reasoning applies in the environmental arena is explored in an insightful and disturbing study by Andrew Szasz, *Shopping Our Way to Safety: How We Changed from Protecting the Environment to Protecting Ourselves*. For people of middle-class means and higher, Szasz states, the tendency has been to improve their environmental surroundings by turning to the market. Instead of working with fellow citizens to promote the common good, those with money are sometimes prone simply to distance themselves from environmental dangers. They move away from air pollution and toxic risks. They install filters on their water systems and shift to bottled water. They shop in organic food stores. In short, they cut themselves off from environmental dangers rather than standing to fight them. Indeed, many show hostility toward regulations that protect the environment or that promote healthier food systems. Such regulations, critics contend (often wrongly), hamper the economic machine that generates their individual income. Regulation

also can overstep government's proper, neutral role while limiting individual liberty.

Szasz has termed this the reverse-quarantine approach. People with
money construct healthy enclaves and healthy supply chains in a landscape that is then allowed to slide down. Manufacturing, mining, and
energy generation are sent far away, overseas if possible, soon followed
by the inevitable high-consumption return wastes. When living within
a healthy enclave it becomes easy to ignore the plight of miners, factory workers, and those who live on or near the resulting wastes. It is
easy to overlook the low-wage workers in impoverished places paid to
dissemble discarded electronics and scavenge reusable materials. Indeed, in the market-age worldview, a person is only responsible for his
own actions, not for the actions of contracting parties. A consumer is
not responsible for the ills associated with making goods or generating
power, nor is he responsible for how his wastes are handled so long
as he pays someone to haul them away. In sum, Szasz reports, a clean
environment for many has simply become another market commodity
allocated according to price, another benefit available for those with
money to spend. The good-citizen way to deal with deterioration is to
fight it; the smart-consumer way is to buy your way out.

Environmental problems arise because we misuse nature. Our actions are shaped by available technology and other material factors.
Human population also plays a vital role. Yet we act toward nature
as we do in critical part due to cultural and cognitive factors, chiefly
those considered in the previous chapters. Our conduct reflects how
we see the world and understand our place in it. It is guided by the
moral frameworks in which we live, by our short time horizons, and
by our sense of being, in essence, rights-bearing autonomous individuals. The primacy we attach to individual liberty and equality, as noted,
plays a critical role. Our long escape from the strictures of communal
duties also fits in prominently. Our daily lives are centered on the market and we define ourselves, as others often see us, predominantly by
our roles as producers (job-holders) and consumers. Within the public
realm, talk and action are constrained by an insistence on objectivity,
leading to overuses of science and scientific methodologies. Nature is
fragmented into private parcels and commodities; it is mere physical
stuff, controlled by owners and valued by market processes. Social and
natural communities have no particular value as such. And other life
forms gain value, if at all, only by their immediate and obvious contributions to human welfare.

Looking around the world, it is clear that we are collectively still guided by a ceaseless urge to dominate nature, to harvest it and improve it, as if nature's ways contained no embedded wisdom. Using the language of reason and often believing what we say, we have been clever in denying even the plain testimony of our senses, clever in claiming that obvious degradation is somehow a good thing, clever in claiming that vice is somehow virtue. In the United States, all of this is worsened by a still-lingering sense that nature's resources are essentially unlimited, if we just work hard enough to find them. American culture was forged in a colonial-era setting in which nature seemed abundant if not limitless, in which there were always fresh lands over the horizon. Abundance gave rise to a cultural tendency to consume, degrade, and move on, a tendency, now market-driven, that remains distinctly evident even as we recognize that the factual truths are otherwise.

The vast bulk of environmental-related research taking place in universities and research centers today, at considerable public expense, is scientific and technical. The science part is undertaken, like all other science, to generate new factual knowledge. The technical and engineering part aims to generate new, greener technologies. Implicit in this work is an assumption that our environmental ills are either caused by a shortage of facts and good technology or will be solved with more facts and better technology. Yet we do not remotely make use of the information and technology that we already possess, and there are reasons to doubt that more facts and better technology will bring much change. Facts and technology bring gains only when they are put to use. In the case of most environmental problems, we already know enough to bring about sizeable improvements using technology in hand. And yet much technology sits on the shelf. We regularly miss chances to apply what we know.

It is telling that university environmental programs are almost always run by scientists and are largely staffed by scientists, typically scientists who study nature. The implication is that our problems reside in nature or that solutions will be found in nature. But our problems are not there. Our problems are in and among people. Human behavior is the cause. Behavioral change is the solution. Bad behavior can stem from ignorance and from a lack of available tools. But we are preeminently cultural beings, guided by worldviews and shaped by senses of value. We will not use the world differently without significant shifts in prevailing culture, in the ways that we see and value the world and understand our place in it. If we want to follow a better path, improving our uses of nature, we must first change our cultural selves.

The Land Community and Its Health

The work of cultural reconstruction can begin by looking to nature and recalling the fundamental truths about it and our dependence on it:

Nature is an exceedingly complex interlaced system involving millions of different species and a dazzling array of geophysical elements. These elements interact according to physical principles and biological processes, variously uniform and chaotic, processes that have evolved over time by fits and starts in ways that increase the diversity and complexity of life and enhance the planet's overall productivity.

Nature's productivity is directly linked to its ecological functioning in systemic terms. We depend upon that productivity for our sustenance and survival. As best we can tell (for instance, from the United Nation's Millennium Assessment), we are overtaxing that productivity in critical ways and degrading the planet's functional capacities. In one much-used phrasing, our overall ecological footprint is already bigger than the planet.

Our knowledge of nature is extensive but still woefully incomplete, and our capacity to gain knowledge through our senses and using reason and technology faces constraints. Even with computers the data about nature are overwhelming in terms of our capacity to bring them together in ways we can grasp.

As humans we are members of social and natural communities as well as individuals possessed of free will. We are defined in vital ways by our communal roles and our surrounding webs of interdependence as much as we are by our separateness.

As for morality and normative value in general, we are free to attribute or create it as we see fit within the constraints of free will; value does not come pre-packaged in the world, waiting for us to find it, and human rights are no exception. All are intellectual tools that help us navigate our ways, individually and collectively, in this world.

If we are to live well in nature over the long term our moral values and normative assessments need to be ones that help us do so: ones that accurately reflect nature's ways and our dependence on it, that take into account our limited knowledge and nature's dynamism, and that recognize the importance of healthy communities as such. It is thus misleading to say, though literally true, that we are free to choose cultural values as we like. Yes, we are largely free to choose, but different choices will lead to widely different outcomes, for people and nature.

The first building block, then, for a revitalized culture is an understanding of nature as an integrated community of life. Aldo Leopold decades ago termed it the "land community," a community of rocks, soils, waters, plants, animals, and people. The land is a community and we are part of it, embedded in it and ultimately dependent upon it; this is the needed point of beginning. Nature is not simply a warehouse of resources and parts, some valuable, most worthless. It is not simply a place we draw upon for inputs for our economic system. It is the overall context in which we live, and everything we do must fit sensibly with in it. Our economy must be a subset of this land community and must operate in ways consistent with its healthy functioning.

It is perhaps not accurate to say that the parts of the land community *cooperate* with one another given that cooperation implies intentional volition and few parts of nature possess it. Still, the term best expresses the kind of evolved patterns of interaction that allow the parts of nature to endure over time and that make the system as a whole function and endure. As ecologist William Ophuls has put it, "If nature could be said to have an ethos, it is mutualism—harmonious cooperation for the greater good of the whole that simultaneously serves the needs of the parts." Evolution, he relates, tends "toward a luxuriance of mutualistic symbioses." As a result, "there is no such thing as an individual life because organisms cannot by themselves sustain life. For their sustenance, they depend completely on the whole system."[3]

In the case of human life, our interdependence with other life forms and natural processes shows up on and in our bodies. They provide home to hundreds or thousands of different microorganisms, on our skin, in our mouths and digestive system. Our health is highly dependent on their presence; we could not live without them, nor in many instances they without us. So hugely varied are the life forms on and within a single human body that environmental philosopher Baird Callicott has taken to referring to an individual person as a superorganism. A human being is a highly integrated community of life forms that operates as a unit, even as it is complexly linked with and dependent upon surrounding life and physical elements.

Cultural reform, then, needs to begin with a new understanding of the nature of existence, with a new ontology. We need a foundational moral understanding that embeds us in webs of interdependence, one that gives as much emphasis to the webs of interdependence and to the community's emergent properties as it does to the moral importance of the autonomous individual human as such. We need to begin, that is, by returning individuals to their natural homes, not losing our senses of individuality but blurring the lines and giving much greater value to the land community as such.

This land community in which humans are embedded can be more or less healthy in terms of its ecological functioning and its ability to withstand stresses and disturbances. This assertion—that communities as such can be healthy—is basically a normative claim, a claim of value that draws upon factual truths about nature but goes beyond those facts, just as all normative claims go beyond facts. Land health is thus not merely a *description* of nature; indeed, scientists would rightly object to it if it were presented that way. And this is so for reasons already covered at length—because it is a normative assertion, and normative values always reach beyond science. Land health is an ecologically based claim of value. And a powerful case can be made that it ought to stand as our preeminent claim of value, more important than the self-evident truths that Jefferson proposed.

The normative goodness of land health reflects the many benefits that we get from healthy lands. It reflects the many ways that well-functioning ecosystems are more valuable for us, generating (in recently popular language) more beneficial services than do ecosystems that function more poorly. But the normative value of land-community health isn't simply about direct benefits to humans. It is about the welfare of the community as such. Land health benefits the community as a whole, including its human members; this is perhaps the better way to present it. In this light, our welfare as a single species is a component of, or derivative of, a welfare that transcends and includes us.

As for the health or welfare of the land community, it is best defined in terms of the general evolutionary trajectory toward greater primary productivity, toward increased interdependence, and toward improved efficiency in the uses and reuses of nutrients. In complex ways other life forms play roles in these ecological and evolutionary processes. The maintenance of land health accordingly requires the presence of these life forms in and among us as valued community members.

We cannot literally say that the land community as such values these functional attributes and evolutionary tendencies. It cannot, because the community is a not a conscious, valuing being. But these attributes of community functioning are hardly ones that we humans have created. They relate to the physical ways nature functions and has functioned over millions of years. They reflect the modes of functioning that directly sustain all life. As valuing beings, as the ones who attribute all moral value in the world, it is entirely appropriate for us to give these facts moral significance. It is appropriate for us in our moral thought to embrace the land community as an organic whole and to value it as such. We can embrace the health of this community as one of our guiding standards of goodness, in the same way that we view human health as good.

For these reasons, the healthy ecological functioning of that community—land health, for short—should become a prime value in our culture. We should choose it as a normative goal that we unceasingly promote. Its maintenance should become a prime focus of governmental efforts. On this point, we simply cannot have governmental neutrality; we cannot leave it to individuals to decide themselves whether they will or will not value the land community. We thus must affirmatively reject former Vice President Cheney's ill-directed claim (and ones like it) that conservation is not a legitimate public policy.

Before bringing into this foundation-building process the next vital element—recognition of our ignorance—it is appropriate to consider briefly how land health as an overall goal relates to our uses of nonrenewable natural resources. Nonrenewable resources are parts of nature that we find valuable, that are limited in supply, and that are generated by nature, in human time-scales, either not at all or in quantities far less than we use. Nonrenewable resources pose challenges when it comes to keeping the earth equally valuable for future generations. Once used, nonrenewable resources are gone except as they can be recycled. Current use then inevitably reduces the supply available for future users. Tradeoffs are necessary, if the resources are to be used at all, and not simply put off-limits forever. How should we make these tradeoffs, consistent with land health as an overriding goal and with due regard for social justice, including justice toward future generations?

For starters, we might recognize that nonrenewable resources are often components of the earth—copper, for instance—that are embedded in the ground and play no roles in ecological functioning. Whether copper is or is not in the ground makes little difference to land health defined (as it is here) in functional terms. Of vastly greater concern to land health are parts of nature that are critical to ecological functioning—topsoil above all—and that are essentially nonrenewable because nature generates them so slowly. These resources, however, are ones that we can use without degrading or consuming them, and this plainly must be our goal. (Particularly topsoil has routinely been misused.) In terms of land health, then, we know what we need to do.

Social justice concerns are not so easily satisfied. Use of a resource today leaves it unavailable for future users. But present-future tradeoffs on resources use are likely less common and less important than we might first guess—or they would be in a culture that valued land health. Many nonrenewable resources are ones that generate grave ecological harms when they are extracted and used. Coal and oil are prime examples. Our use of these resources thus clashes with land health. For that reason alone we should stop using them, without any need to rely on a moral duty to keep them around for future generations. That still leaves us dependent on some resources that are nonrenewable. In the case of these, the best approach is simply to use them sparingly, recycle them fully, and push hard to find substitutes, particularly renewable ones.

Ignorance and Nature's Ways

Inevitably our understanding of this land community—our understanding, for instance, of the functional roles played by various species—is quite limited and likely will remain that way. We often talk about decision-making that takes place under conditions of uncertainty. Often more accurate and honest is a phrasing that openly confesses ignorance. Yes, in some settings we are knowledgeable but not quite certain of underlying facts. But in many others our knowledge is much less than that. Often we know very little, and can't even gauge the depths of our ignorance.

This ignorance (a recognition of it, that is) needs to play significant roles in our cultural values. It can do so by prompting us to act more cautiously with nature, by encouraging slow, well-considered actions that anticipate surprises and leave ample room for mid-course corrections. A confession of ignorance encourages us to avoid placing bets we can't afford to lose. It also means new burdens of proof to use in evaluating possible harms, burdens that err toward safety. The choice of a burden of proof, as already considered, is a normative question of importance, one that should be addressed thoughtfully and that, in dealings with nature, is very much public business. The same can be said about the types of evidence that should be heard and weighed when assessing claims of danger.

Much good conservation writing has centered on the inevitable incompleteness of our knowledge and how we might accept it. Agricultural pioneer Wes Jackson has explored the benefits of an "ignorance-based" approach to education, action, and governance. We are vastly more ignorant than we are knowledgeable, he teases, so we ought to go with our strength. We play to our ignorance, Jackson says, when we pay close attention to the way nature functions in a given setting and try to learn from it. What ways has nature evolved to inhabit a given landscape, with its particular climate, hydrology, terrain, and soils? How have life forms evolved so as to flourish under particular local conditions and with particular local disturbance regimes? The closer our modes of living mimic those of native species, the more likely they are to take advantage of nature's embedded wisdom, even when we can't really perceive that wisdom ourselves.

Our guiding principle, Jackson urges, should be to use nature as our measure. We should seek out and put to use lessons that have arisen through local evolutionary and ecological pressures. The closer we stay to nature's choices—using local species, retaining maximum local diversity, respecting natural hydrological systems—the more likely our ways of living will endure under local circumstances. Jackson's own much-cited work has involved research into new ways of farming, including new crops. He and his colleagues seek to grow food in the former tallgrass prairie using crops and methods that imitate the functional composition of the former prairie. This means, he says, practicing farming that relies on perennials rather than annuals, that uses a mixture of species rather than a monoculture, and that includes species that fix nitrogen in the soil to complement species that lack that capacity.

Nature-as-measure is the kind of cultural principle that could usefully apply widely. It deserves prominence. Forestry practices can mimic natural forests, using selective logging and mixed species while taking steps to maintain functionally important native species (often predators that keep herbivores in check). Nature's ways of using semi-arid grasslands are similarly worth imitating, both in terms of grazing patterns and in the use of grazing animals that do not degrade waterways or require special protections against predators and pests. Lands should be used in ways for which they are ecologically well suited without material alteration—a line of thinking that will require major shifts in the presumed development rights of private landowners (an issue in the next chapter). Inevitably we need cities as places for large numbers of people to live; indeed, large cities can make good ecological sense when they significantly lighten the human presence in rural areas. Urban areas cannot remain ecologically healthy in any full sense of the concept of health. But they can be designed and reformed with specific reference to their effects on ecological processes—on water flows in particular.

Central to a new, ecologically grounded culture should be a long-term perspective, a recognition of the moral value of anticipating the needs of future generations and not curtailing their options or aggravating their burdens. The popular notion of sustainability gets at this point. As commonly interpreted, however, sustainability does not entail distinct limits on our uses of nature—on the sprawl of cities into the countryside, for instance. Instead it calls simply for more careful, planned expansion. But geographic expansion cannot go on indefinitely. It is not enough simply to consume irreplaceable nature more slowly. At some point, limits must be put in place, clear lines drawn and respected. And they need to be put in place, many of them, quite soon.

Wildness and Integrity

So far the discussion about respecting nature has spoken about the health of the land community in functional terms. Ecological health is an overall norm that should guide and constrain our uses of the landscapes we inhabit and the places where we get our food and resources. For many sound reasons, we are wise to use certain other lands far more lightly than this, and to set management goals for them that more closely mimic natural conditions. We should maintain certain places here and there—well-chosen places—in wilderness-like conditions, even as we admit that we have already altered all parts of the planet at least slightly and indirectly.

Critics might cry that wilderness is an artificial construct, a product of the human imagination, or that it is a conceit of wealthy people who do not have to go out and work the land. The reasoning involved in such claims is plain enough: It is far easier, yes indeed, to talk about leaving lands untouched when one draws an ample salary and is well fed. Even so, these criticisms of wild-lands preservation are all badly aimed. Of course the term "wilderness" is a human creation, as is the underlying idea. Of course (as some point out) early tribal peoples living in nature may have had no word in their vocabularies that referred to places in nature untouched by people (though likely they had words that referred to places that were spiritually charged). But these observations are diversionary and pointless. All words, all ideas, are human constructs. Given our plight we very much need words and ways to talk about lands only slightly affected by people, just as we need ways of talking about lands that have been extensively altered.

Areas maintained in wild condition—call them what we like—are essential to the long-term quest to live well on land. Their benefits have been catalogued again and again. Many benefits have to do with the maintenance of genetic diversity. Others have to do with the value of wilderness areas as places to study nature's functioning, as ecological test-plots to help us see how we have altered other places and with what consequences. In his well-known final essay on wilderness, Aldo Leopold somberly anticipated that modern America's attempt at the oldest task would end badly, particularly in semi-arid lands and other ecologically challenging places. They would end badly, much as earlier human efforts to endure in sensitive places already had. When that happened, Leopold urged, we would need to return to wilderness in search of a "more durable scale of values." By that he meant a scale of values, drawn from the careful study of wilderness, which recognizes and respects nature's ways of functioning in a given place.

Wild areas are repositories of evolved natural wisdom. They provide home to life forms that cannot thrive in and among people. Even as enclaves they can sustain ecological processes that benefit surrounding, human-occupied lands. Not incidentally, they also provide places for human recreation and awakening. In Leopold's scheme of wilderness values, wilderness was needed most of all as a tool to help society gain virtue, particularly the virtues of humility and restraint. Wilderness preservation could play a key role in bringing about cultural transformation. Indeed, so central to Leopold was the role of wilderness in cultural change that he had doubts about land-protection efforts that did not have, as a central aim, the use of the lands to foster such change.

The literature on wild-lands protection, much of it penned by scientists, explores the standards that might best serve as management guides. Here we need only take note of the conclusions. The common sentiment is that wild lands should be managed, insofar as possible, so as to sustain the full range of species that lived on the lands at some point in the past, perhaps before industrialization came along, perhaps before European settlement arrived, perhaps even further back in time. Whichever time is picked, the normative claim is that we should retain as many then-resident species as possible and do so in numbers that approximate their then-existing populations. This normative goal is often talked about as the land's native *integrity*. It is a goal focused more on biological composition—what species are present, and how many—rather than, as the goal of land health is focused, on a land-scape's ecological functioning. Species composition and ecological functioning, though, are closely tied. It isn't possible to maintain the full panoply of species in a place without also maintaining something close to then-prevailing modes of ecological functioning. Integrity as a goal would similarly require the protection of biotic communities as such, not simply the protection of species one by one. In short, integrity has to do with the maintenance to the extent possible of natural conditions as of a particular time and taking into account the kinds of dynamic changes that would likely have unfolded even without any human presence. It should not need saying that, when offered as a value goal, integrity is a normative choice, not merely a scientific description.

Looking, then, at landscapes as a whole, a sensible overall goal is this: to maintain the ecological health (land health) in functional terms of all lands, and to go further in designated patches of wildness to get as close as possible to sustaining their integrity. Both parts of the goal, of course, need more sustained explanation and inquiry than they have here, and they have received it. Of course our knowledge of the underlying science remains limited, yet land health and integrity are nonetheless clear enough to stand as highly useful goals. They are clear enough to serve as shared normative ideals, on their way to becoming, we should hope, new self-evident truths.

Other Species

The mention of land health and integrity draws attention to the roles of other life forms, more than eight million of them according to recent guesses. In the recent past, educated estimates about species numbers have varied from as low as three million to as high as 30 million, with a few estimates reaching to 100 million. To date taxonomists have identified and given names to some two million of them, with more turning up with every collection effort. Any overall number involves estimation. The challenge of counting species is heightened by a certain arbitrariness in deciding what we mean by a species and which life forms we count. Estimates of species diversity routinely exclude bacteria and certain other single-celled forms of life. The challenge of counting them would be truly immense. It is, it turns out, a serious research challenge simply to categorize and count the bacteria that live on and in the human body. That effort, the human biome project, is likely to go on for years. A major research effort is needed simply to take a spoonful of sand from a beach and identify the many types of microorganisms contained in it.

Looking ahead, looking toward a (literally) revitalized culture, we need to show greater respect for other life forms than we have. That much is clear. But for what reasons and to what extent?

A small proportion of all species are directly valuable to us and are rightly respected for that reason. Vastly more species are valuable in that they help sustain ecological processes and thus land health. We benefit from them in this indirect but essential way. As for which ones are functionally valuable our knowledge is highly incomplete, even in landscapes that have been long-studied. Presumably many are not, but it is difficult to say with any high degree of probability. A common normative claim is that we should keep all species around since we never know when a species might become valuable, either because we learn something new about it or because our circumstances shift. Many species could supply useful genetic material in future plant or even animal breeding. Others are useful simply because we enjoy seeing them or being around them, in zoos or in the wild. The direct and indirect benefits to us are many.

A central question in environmental philosophy has long centered on claims that other life forms, as individual animals or species, have or ought to have intrinsic value that we should respect, value that is unrelated to any benefit we receive from them. They have or should have value in and of themselves, without regard for how they affect us.

This normative claim is easier to engage when we realize that intrinsic value simply means (as noted earlier) value that humans attribute to a thing apart from any contribution it might make to human welfare (other than contributions to our sense of being virtuous). It does not mean—or certainly need not mean—value that exists apart from any human recognition or attribution of it. As for this latter kind of intrinsic value, many people have embraced it and still do. They embrace the idea that nature has a kind of value that existed before humans came along and will continue existing even after humans disappear. Such reasoning fits easily in many religious schemes (value put in place by God, not people). It is certainly not a line of thinking that can be rejected as nonsense. But the practical reality is that such intrinsic value, not arising by human choice, still requires humans to identify it and in some way legitimate it. It somehow has to enter into our consciousness if we are to take it into account. There thus remains a need for humans to decide consciously that a particular creature or species has value, whether they think they are creating the value themselves (a common philosophic stance) or believe instead that it exists apart from them (a common religious stance). Human decision is always needed, and the two types of decisions—attributing value and recognizing pre-existing value—do not differ much.

Many philosophers have resisted the idea that intrinsic value can exist in a category or intangible, such as a species or a biotic community or ecosystem. Their resistance arises because species and communities are not distinct physical things and value, they assert axiomatically, is an attribute of a thing, perhaps only a living thing. It is not clear, though, why this lack of physical thingness should be an objection to intrinsic value. We routinely value ideas and forms of knowledge, apart from any embodiment of them. A Shakespeare play is valuable as a literary composition apart from any printing of it; for many, the character Harry Potter is valuable even though entirely fictional. A species may be a human-created category, a mental concept, but it is a concept that refers to real patterns among physical creatures; it is a way of talking about real, physical differences in the world much as gravity is a real physical force. Biotic communities also have real existence in that they are unique combinations of living creatures and physical elements that interact in special ways. To deny that a biotic community can have value apart from its living components is much like saying that a great painting has no value apart from the oil paint and canvas used to create it.

It is thus entirely sensible to attribute value to species and communities as such, just as it is sensible to attribute value to other living creatures and to special physical places and features in landscapes. From many perspectives, an earnest, collective desire to protect all life forms is a morally worthy and honorable stance. It is a moral position that reflects a grasp of the limits on human knowledge, particularly on our ability to distinguish valuable life forms from expendable ones. It also reflects a virtuous willingness to act cautiously, leaving room to correct our inevitable mistakes. The disappearance of a species in a landscape is often (though not always) a signal of ecological change that diminishes the landscape's functioning and is morally troubling for that reason. Species loss highlights the fact that change is taking place and that a landscape is consequently less suitable for at least one life form. In the religious mind, all forms of life might possess value due to their divine origins and have value for that reason. Value might also be attributed to species and communities as a way of highlighting the importance of protecting them so future human generations can enjoy them. In short, the reasons for protecting all species are numerous and, when combined, quite potent.

Individual Creatures

Far easier for many people is to sense that other life forms have moral value as individual living creatures, particularly mammals with high mental functioning. The focus here is on individual beings as such, apart from the instrumental roles they may play and the benefits they confer in particular ecosystems. This moral reasoning is mostly aimed at animals, rarely at individual plants. Typically it is strongest in the case of animals that do not live in the wild—companion animals, zoo animals, and domesticated animals generally. People have removed these animals from any natural home and typically bred or raised them so that, quite often, the animals could hardly survive in the wild. Having stripped them of the chance to live in the wild we take responsibility for their fates.

Much animal-welfare sentiment has risen up independently of any concern about ecological degradation generally. That is, concerns about environmental decline and those about animal welfare often exist separately. At quick glance the two concerns might seem closely joined if not overlapping, and they can be. But concern for individual animals as such can also collide with concerns about ecological systems and this collision needs to be understood.

The health of a land community requires a mix of species interacting in healthy ways and with populations of these resident species limited to functionally reasonable levels (not too low or high). When humans intervene in natural systems—as we have, essentially everywhere— they trigger changes to these mixes of species and population levels. Often, human intervention causes some populations to rise far above the numbers they would have otherwise had. Such species can become pests in the sense that they alter their home landscapes in ways that bring declines for many other species—other forms of life that are also presumably valuable—and declines also in productivity and functioning. How should we respond when a handful of species explode in numbers like this? Should we think about the health of the land community as such or (instead or in addition) pay attention to the value and welfare of each animal individually?

This question produces different answers from people, reflecting differing normative stances. The perspectives can particularly clash in clear ways when the call goes out to cull particular populations by hunting or through other mortal methods as a way of reducing excessive numbers. Deliberate killing can certainly seem inconsistent with even a modest moral concern for the creatures being killed. On the other side, when people have indirectly caused an animal population to irrupt, harming other species and perhaps land functioning, then people have brought on the harms themselves and might feel morally responsible for them. In this light, the effort to control one population (by hunting, trapping, poisoning, or introducing predators) is, in effect, an effort to curtail the harms being imposed indirectly on other species. Hunting an overly abundant species therefore diminishes the overall human impact on nature, rather than expand it.

The claim that individual, nonhuman animals ought to have moral value, elevating them above rocks (where Descartes famously put them), is an argument that has long encountered a stiff headwind. One reason is likely a concern about what it would mean to attribute value in this way. Animal advocates typically argue their case for expanded moral coverage by using a particular mode of reasoning, one that draws upon Western moral thought after its turn toward individualism. The point of beginning for the pro-animal argument is that humans are morally worthy creatures because of some attribute or capacity that they possess. If this is so, then it makes logical sense (consistent with the principle of equality) to conclude that nonhuman creatures should possess moral value if and when they also possess this same value-creating attribute or capacity. The argument, that is, is basically deductive in form. Attribute or capacity A is the source of moral value (a general normative claim); certain nonhuman animals along with humans possess A (specific factual claim), which means these other creatures should also possess moral value (deductive conclusion). The reasoning is sound, and thus the conclusion, so long as the key normative premise—that moral value arises from A—is correct. Of course most if not all animal-welfare advocates have strong sentiments about animals and are likely motivated by these sentiments. But they are nonetheless inclined, when presenting their arguments fully (academics above all), to steer clear of sentiments and preferences and to respect the cult of public objectivity.

This reasoning, predictably, has led to speculation and disagreement about the all-important attribute or capacity A that allegedly gives rise to moral value. What is it that makes humans morally special? A long-time Judeo-Christian answer was that humans were special because they were created in the image of God. Other species, it was thought, were specifically created to serve human needs in some way, if only to provide illustrations of how people should and should not live. In the medieval era it was often said humans were different because they possessed religion or could speak and reason. By the seventeenth century, reasoning tended to focus on the presence or absence of a soul; this was the defining human feature. Because other species (it was said) lacked souls, they were no different morally from other complex, noise-making machines.

Recent literature, addressing this age-old issue, has tended to focus instead on neurological functioning, ignoring earlier religious reasoning and any reference to souls. In one view, the key value-creating attribute is simply the ability to feel pain and thus to suffer, the point of view embraced by Jeremy Bentham whose moral reasoning gave primacy of place to pleasure and pain. Other writers point instead to consciousness or some similar high form of mental functioning as the key attribute. More narrowly drawn is the claim that moral value resides in self-awareness and an ability to anticipate the future and chart a life course. A few writers have contended that we should avoid drawing a single line between the moral haves and have-nots. Instead, moral value might rise up by stages based on several specific capabilities, leading not to two distinct moral categories (moral haves and have-nots) but to gradations of moral worth. In all of this reasoning, the guiding idea, again, is that if the morality-giving trait can be agreed upon, then other creatures that share the trait would rise up in moral terms.

This reasoning, it should be clear, is very much an expression of the current confusion about morality and its bases. It reflects also the modern tendency in public affairs to push normative issues aside and speak in objective terms. Objectively speaking, an argument can be made in favor of certain animals if it begins with an agreed-upon source of moral value—hence the search for such an all-important, moral-vesting attribute or capacity. The pro-animal argument thereby doesn't have to start by proposing a new source of morality, which it couldn't do in any nonsubjective way.

A far more likely explanation of our current moral worldview, however, is that humans have moral value, not due to any elusive attribute A (based on brain functioning or otherwise), but simply because we are human. We have value, that is, because we think and say that we do, because we have decided that we do. It is a stance we embrace as a moral normative axiom, a self-evident truth. It is not a conclusion we have reached using facts and inductive reasoning. (The claim that high mental functioning [a fact] creates moral value [a normative stance] in any case looks like a questionable jump from is to ought.)

Over the millennia, as noted, peoples have largely embraced a different, more limited axiom. They have typically confined moral value to people like themselves or to people within their group: to "our" people or the chosen people, and not others. Slowly, painfully, this moral axiom has given way to a new one in which all humans have value. It is this history of extension in the moral community that animal-welfare advocates draw upon when they push to expand the community beyond the species line. But it is not clear that this can happen given the many differences between people and other species. More important, it is not clear either that the argument is needed or that it is strategically prudent.

Driving the long expansion of moral value to all humans has been the claim that differences among people are not morally important, their races, religions, ethnicity, sex, nationality, and so on. Humans have value—that is the basic moral view—and that should mean all humans, not some artificially bounded subset of them. Expansion did take place, the circle of morally worthy people did expand. But there is little sense in this history of expansion that humans possess value because (and only because) of some particular capability that they display, some specific difference between them and other species. Rather, it has unfolded because of the lack of any morally meaningful differences among people themselves. People reside within the moral community circle (once a small community, now global) simply by reason of their humanity, as Christian scriptures affirmed long ago.

Animal-welfare writing in a sense simply wants to toss out this axiom and replace it with a different, new axiom. It would start not with the axiom that all humans have moral value, but with the claim that some larger group of creatures has value based on some particular identified factor. There is nothing at all illegitimate about this approach. But the chosen reasoning nonetheless reflects our current confusion about moral principles, where they come from, and what makes them legitimate—topics all covered earlier (chapter 3). Moral values, as we've seen, emerge out of deep-seated moral sentiments mixed together in the formative stage with facts about the world and our powers to reason. Our collective moral sense today is that people have moral value, and we act on that sense. We could, in exactly the same way, embrace the stance that other creatures also have moral value. And we could do so simply because we feel that this is so, because our deep-seated sense is that humans are not all that different, as Darwin long ago told us. Facts and logic might well help in this effort. But they are not really needed. It is entirely legitimate to press the pro-animal moral stance directly and emotionally and to invite others to join in embracing it. We are, to repeat, value-creating beings.

In all likelihood, the appeal of the animal-welfare stance is mostly based in sentiment. That being so, it is diversionary if not confusing to offer arguments that are so strictly logical and fact-based. To frame the argument in this way is to ignore the foundation of morality in sentiment. Indeed, it is implicitly to discredit the legitimacy of such sentiment. The message embedded in it is that morality exists only when it can be supported, if not proved, by facts and reason. Yet this simply isn't true. Moreover, it is a message that, overall, is quite hostile to expansions of moral value, whether the expansion is to other living creatures as such, to species or to entire communities. The logic-based argument might work for many audiences in the case of moral value for high-functioning primates; such animals are perhaps similar enough to people for the argument to work. But it isn't likely to work in gaining moral standing for communities, wild rivers, and special landscapes—all of which are too different from humans for any extensionist argument to help. In the meantime, the noisy disagreement among animal-welfare advocates, as they search for the elusive attribute A, makes the whole intellectual effort seem inconclusive if not arbitrary. Why should moral value rest on one capacity rather than another; on one fact rather than another? There is no good answer because morality, as we should know, doesn't arise from facts alone. Reason alone is not adequate to select among the many proposed attributes, nor is it, in truth, adequate to show that the humans-only moral view is somehow bad.

What we need to see more clearly is that value comes through a process of collective choice. Sentiment is a central guide for that choice, probably the most important one. When we see this then we can more readily also see that the value we attribute to other living creatures need not be the same as the value of humans. It can be moral value of a different kind, just as moral value attributed to a species or biotic community can be quite different. It is, accordingly, wrong to assert (as critics sometimes do) that if animals have moral value (even rights) then they must be put on the same level as humans. Not so. Similarly, it is wrong to assert that the case for animal welfare fails because it has not been proven, or that it rests on logic that isn't airtight. Proof and logic have nothing to do with it. Moral sentiments do.

The reasons for respecting individual wild creatures—beyond the direct benefits to us already mentioned—can be quickly mentioned. They overlap with the reasons for protecting species and biotic communities. Other creatures play important ecological roles, often ones that we dimly understand. We also have good cultural reasons for showing them more respect, without regard for any desires to do so. Efforts to protect other creatures can help foster the cultural values of restraint and humility. They can lead us to alter nature more gently, in ways that yield multiple benefits. Beyond that, they can help instill senses of awe toward nature and toward the capabilities of other creatures, enhancing our sense of being participants in a larger community of life as distant kin of all other life forms.

Rooting Culture in Nature

The points of beginning proposed in this chapter together invite a significant turn in the course of Western culture. They pose, in particular, a challenge to the dominance of liberal individualism in its political guises all across the political spectrum. The form of right-living that they project is essentially a communitarian one, not a liberal one. As a moral stance it aligns better with the true temperament, evolutionary breeding, and physical needs of humans, especially as social beings. The lone individual can get ahead and perhaps thrive through aggressive, self-centered competition. The species as a whole cannot.

Front and center in a better-grounded culture is the land community. It is the big picture of which all else it is a part, our source of sustenance and inescapable home. This land community of over eight million species is complex far beyond our understanding and its productivity depends upon this complexity. It turns, not just on its great variety in life forms, but on their evolved ways of interacting that have given rise to capacities much greater and different than the capacities of the parts in isolation. Life as we know it would grind to a halt if nature were completely fragmented, if it were turned into the kind of resource warehouse we seem to think it is.

Nature is not a stockpile of resources. Nor is it well understood as *flows* of resources (the early twentieth century view) or as flows of ecosystem services (the twenty-first-century revision). It is an integrated community of life of which we are a part, an overall whole in which the components cannot be understood except in relation to the whole. Well tended it is a fine place for human life. Well tended, that is, it is the kind of natural home we have been designed by evolution to inhabit. This land community can be more or less healthy in its functioning. Particular landscapes can also be more of less whole in their integrity. Our success at our oldest task requires us to maintain this overall health. It requires also that we retain diverse landscapes that display here and there something close to the integrity they had centuries ago. By all appearances, these goals in turn require major changes in our behavior, which means, at root, major changes in modern culture.

The trajectory of modern liberal culture (Western culture particularly) is one that has entailed a good deal of line drawing. The embrace of a new, ecological cultural stance will involve blurring many of these lines, even erasing them here and there. In a common formulation, modernity is a worldview in which time and life are linear and progressive rather than circular, in which lines are drawn between past, present, and future. Renaissance humanism, extended by Enlightenment thought, vested humankind with greater moral value, distinguishing it more clearly from other life and celebrating its unique powers. That sharper line, between humans and other life, soon gave rise to further line-drawing among individuals and to an unprecedented emphasis on the individual human as such. Mind drew apart from matter; reason drew apart from sentiment; logic drew apart from intuition—not completely in any case but in theory and in many ways in practice. The physical matter of the world was divided from realms of spirit; divinities increasingly were transcendent, no longer immanent. For people, the personal realm pulled further from the public realm, exposing yet another fault line. Except on issues of public safety and a few key public services, religious and moral values were pushed onto one side of that dividing line, the private side, while objectivity was left to rule on the other side, the side increasingly dominated by science and the science-garbed discipline of economics.

All of this line-drawing, we must see, came with great benefits and also with great costs. Looking ahead, the quest is to retain these benefits as best we can while greatly reducing their associated costs. Inevitably this will mean reform efforts that blur all of these lines. The blurring can't be wholesale, and certainly not haphazard. What's needed is blurring based on sound ecological knowledge and on the kind of deep-seated sentiments that can give rise to new understandings of morality, to moral orders that reflect our lives as members of social and ecological communities, and that call us to higher moral planes in our dealings with other life and future generations.

In the age-old story man took a bite of the apple in Eden and gained special knowledge of good and evil or some power over it. It's a story worth taking seriously for it captures our earthly plight. We do, in a sense, have the power to define right and wrong, good and evil. But consequences flow from our choices, and we shall live with and be responsible for them.

Social Justice

The line between the use and abuse of nature is one that needs to reflect more than just a solid understanding of nature and our dependence on it, and more too than a recognition of our limited knowledge. A major part of this intellectual line-drawing has to do with the human social side of our existence, with our embeddedness in the community of human life. Social justice, needless to say, is a sizeable topic. Much of it has little to do with nature. But good-sized components of the social agenda do relate to nature and our uses and misuses of it. For various reasons, it is hard to imagine a people collectively living on land in good ways unless they treat one another fairly. Social justice is thus instrumentally important in the work to achieve healthy lands. But it is more than that, for it enters into how we would define the goal of good land use itself.

Justice includes consideration of how we share the earth and its resources with one another, in a territorial sense and in our relative rates of consumption. It has to do with what happens to our pollution and other wastes and who ends up living in degraded settings. It has to do with governing power and with the ability of local people to manage their homes in ways consistent with their long-term health and reflective of their moral aspirations.

Finally and critically, the social-justice side of things raises pointed questions about private property rights in nature—the body of law that, perhaps more than any other, organizes managerial power over nature. We take private property for granted, and should not. As land- and resource-scarcities expand, as more and more wealth is transferred to those who claim private rights in nature, the institution is becoming less and less legitimate. Private ownership is a morally complex social institution ultimately grounded in state-supported coercion. When well tailored it can bring vast gains, widely spread among people. Poorly tailored or allowed to get out of date it can become what it often has been and in many places is, a tool of social domination and exploitation and a shield to ward off calls to act responsibly.

American writing on environmental justice typically defines the subject far too narrowly, reflecting, as we might expect, the unduly limited scope of public normative discussion. In the common American view, environmental justice is chiefly about the racial and ethnic characteristics of the individuals who live near unwanted local land uses (pollution sources typically) and who are exposed to their hazards. Thus limited the problem is basically one of inequality. But the subject first and foremost is about access to the good parts of nature, to productive farm lands, water supplies, forests, fisheries, and mineral resources. It is about managerial power over nature and how the gross proceeds from farming, forestry, mining, and the like get divided up. And it is about sharing access to inherently communal resources, above all to oceans and the atmosphere's limited capacity to absorb climate-changing gases.

Eight Observations

The topic of this chapter is so vast that it necessarily can be covered only with a fast-moving overview. An overview, though, is enough to give a sense of the field—to make sense of this part of our oldest task—and to shed light on how this component of good land use might fit into a larger vision of people living rightly in nature. As a beginning and before taking up the key social justice principles it is useful to set forth eight or so major observations that provide a useful frame.

First, the topic of justice can reasonably be defined so that it includes our dealings with future generations and with nonhuman life. When it is, the topic becomes even more central to the challenge of defining good land use. Indeed, a number of prominent philosophers writing on humans and nature build their entire normative structures on felt duties to tend nature for future generations or felt duties to respect the moral value of other life. Here these matters will be put to one side. Nonhuman life appeared in the last chapter; the moral status of future generations has come up several times and will return in the concluding chapter.

Second, natural conditions around the world vary hugely in terms of their natural attributes and elements and their ability to sustain human life. Good climate is critical, as are elevation, terrain, rainfall (quantity, timing, and reliability), topsoil, and the presence or absence of significant uncontrolled diseases. When it comes to sharing land in a fair way, it isn't enough simply to look at acres per person. A fair-share allocation system would take such differences into account. It is also the case that peoples lacking suitable places to live can end up on lands poorly suited for intensive human occupation. Merely seeking to survive they can use such places in degrading ways for lack of real alternatives.

A related reality is the commonplace truth that uses of nature by people in a given place can readily affect other people and their activities in many ways, in addition to the mere fact that occupation of a given landscape by some people means that others cannot use it. Many land- and resource-use patterns in one place have consequences that spread widely: Air pollution crosses boundaries and borders, sometimes traveling thousands of miles. Climate-changing gases emitted anywhere affect the planet everywhere. River corridors supply particularly stark examples: Dams alter natural water flow regimes, changing the quantity, quality, and timing of downstream flows, disrupting siltation patterns, and blocking migrations of aquatic species. Water pollution has grave effects, most plainly when it leads through eutrophication to dead zones in estuaries and to polluted water supplies for people, livestock, and agriculture. Drainage practices (for instance, subsurface tiling) and levee building have similar ecological effects, often aggravating flooding, promoting artificial droughts, and degrading aquatic communities (especially species that use floodplains for key life events). Interferences with natural disturbance regimes can also go on the list. So can disruptions of wildlife migrations.

Third, the world is characterized by massive inequality among people in relative access to nature and in income and wealth, among individuals and among nations. In the case of much of this inequality, nature plays a sizeable role. People who live in fertile lands rich in resources often have had higher incomes and become wealthier than people who do not, a point illustrated in Jared Diamond's popular study, *Guns, Germs, and Steel*. Within societies and nations, those who claim private rights in lands and resources routinely use their rights to divert income to themselves, often year after year and without the owners lifting a finger (in the process amassing further capital leading, often, to yet greater inequality, in a process charted in Thomas Picketty's bestseller, *Capital in the Twenty-First Century*). Inequality is worsening in much of the world, particularly in developed countries with more mature economies, and the growth in national income largely goes to those who are already the highest-earning and wealthiest. Among nations and different peoples inequality is even starker.

This pervasive inequality in income and wealth has important implications. Poor people struggling to survive are often driven to misuse nature to do so, to overuse soils, waters, and wild animals and otherwise exploit nature to gain basic necessities. For the poor, long-term horizons and ethical bearings are often unaffordable luxuries. The wealthy must share the blame when this happens. Continuing and rising inequality also means that development goals cannot be met simply by growing national economies, particularly in mature economies where growth simply adds more to those who already have much. The same is true of goals related to lifting up the poorest in wealthy countries; again, national growth and increased overall consumption will do little. The solution lies chiefly in a different distribution of national incomes, not greater growth in it. To see this is to realize that, in much of the world, development goals and environmental conservation are less in conflict than commonly thought. If the only real solution to inequality is greater sharing, then there is little need to degrade nature at a faster pace.

Fourth, an often-overlooked reality is that the control over parts of nature commonly brings with it some control over the people who need that nature in order to thrive or even survive. The truth is long-standing: As C. S. Lewis put it over a half-century ago, "man's power over Nature means the power of some men over other men with Nature as the instrument."[1] Those who control water in a dry region have power over those who need it. Those who control the fertile lands can insist that their tenants divide the income with them, if not also pay homage in political, social, and other ways. World history is studded with examples of powerful people taking over valuable parts of nature and either dominating existing users or pushing them away. As the powerful exert control over the good places the losers may migrate to less hospitable places: to disease-ridden swamps or floodplains, to harsh climates, to thin and dry soils and arid lands, and to garbage dumps and pollution hotspots.

This fourth point about nature as instrument of control and exploitation ties to the related point that legal power to control nature—whether in the form of private property rights or territorial sovereignty—exists only when backed by some form of dominating power. To put up no trespassing signs or border fences, arresting those who ignore them, is to use collective power (police, even armies) to restrain the liberties of other people. As noted in an earlier chapter, the liberty gained by a landowner due to the recognition of her private rights in land is matched by the loss of liberty by other people who want to use the same land. It is common in the United States for people to think of private property as some form of private power that exists independently of the state; to assume that the state's role is simply to respect and protect property that arises or continues to exist due to forces independent of law and state power. To the contrary, private property is a creation of coercive law in some form (whether or not written and whether formal or informal). Landowners do not want the state to stay away. Instead, they want the state close at hand; they want police nearby to seize anyone who violates their property rights.

In a moral order that exalts individual liberty, private property is in fact distinctly problematic. Yes, private ownership expands the liberty of owners, but it does so only by restricting the liberties of other people. The same is true of territorial boundaries of political units. Those within a bounded territory enjoy heightened control within their borders, but they gain control by restricting the liberties of outsiders who would share their lands and resources. These realities of property rights and territorial control draw them into the realm of environmental justice. Justice among people vis-à-vis nature calls for a critical look at these boundaries and property claims. Today's property regimes and sovereignty arrangements are far from morally neutral in terms of social justice and ecological effects. Change is in order.

The sixth beginning observation for an inquiry into environmental justice looks to the market, which is often used (as are property rights and sovereign boundaries) as a basis or rationale for warding off claims for greater accountability in the uses of nature. Mythology notwithstanding, market forces for many reasons do not provide anything like adequate incentive for market participants to take good care of nature. Many reasons relate to what are termed "market imperfections"—to failures of legal regimes to internalize external harms, for instance. But the list of reasons is much longer. The market encourages owners to discount the future; in some settings (publicly owned corporations) it seems to force them to do so, in ways that promote land misuse. More than that, there is simply money to be made cutting into nature and selling its parts, often more money, for the owner at least, than comes from keeping healthy nature intact.

A sound reform agenda will doubtless call for significant limits on market activities and for greater economic democracy (as explored in the next chapter). What needs putting on the table here is the fact that markets operate within institutional arrangements and legal regimes and they can flourish within quite a wide range of differing arrangements and regimes. Their environmental consequences, in turn, can vary quite significantly based on these different arrangements, as can their environmental-justice consequences. When poorly designed and arranged the rules governing markets can foster land- and resource-use practices that worsen social justice and thus aggravate abuses of nature. It is simply not the case, not even close to the case, that market forces inevitably prompt those with secure rights to nature to make good use of them: to succeed at the oldest task. (The first US scholarly paper on environmental economics, in 1913, was a study of why this was so.) Whether they even come close to doing so depends on the particular institutional and legal frameworks in which they operate. Those frameworks are human constructs and subject to change. Environmental reform efforts need to include changes to them.

Point 7 brings in the matter of local knowledge and local bonds and the desirability of ensuring that the people who inhabit a place have the power to use local nature responsibly, free of pressures that force that them to misuse it. The good use of nature almost always requires considerable knowledge about the local nature and the history of patterns of human use. As Kentucky writer and farmer Wendell Berry often notes, it can take decades if not generations for people living in a place, particularly on ecologically sensitive lands, to gain adequate knowledge on how to use local lands so as to sustain their long-term fertility. Land-use errors are easy to make and sometimes quite costly, as when soil washes away or livestock degrades plant communities. Book learning is valuable, but not a substitute for hard-earned lessons based on the peculiar features of particular ecological systems.

To use land well people typically need secure rights in them. It is hard to care for the long-term fertility of land when a person's tenure is insecure in any way, legally, politically, or economically. With secure long-term tenure and with reasonable hopes of passing lands on to later generations a land-user is more likely to forge in his home place the kind of emotional bonds that seem to nourish good land use. In any event, good land use needs to be a legally and economically viable option. Local decision-making and control hardly assure that land uses will be good ones. But a lack of local control, much like a lack of long-term tenure, makes good land use far less likely.

These realities, even vaguely sketched, highlight that the allocation of power and land-use rights among people can have significant effects on the land itself, on ecological systems and biotic communities, because they affect how nature gets used. That is, socially just relations among people, based on fair sharing and respect, might well be essential to support patterns of using land and natural resources in ways that stay on the use side of the critical line and don't become abusive. In the view of some, private ownership is close to essential; only in that way can a user invest himself with the land, emotionally and intellectually, so as to make good use of it. This is perhaps not always true. What seems more clearly right, and uniformly so, is that those who make decisions about using nature must be guided by a clear vision of good land use, and they must have the capacities, knowledge, and opportunities to achieve that vision. It is for this reason that an outside power that pushes a hazardous waste dump or polluting facility onto a local community, degrading it ecologically, acts unjustly without regard for the race or ethnicity of the people living there. If invidious discrimination is involved the injustice increases. But the deprivation of local control, sapping the ability of local people to use their lands well, is itself a form of injustice. This form, and indeed pretty much any form of injustice, can readily disrupt or derail efforts by local people to do good work.

These days, particularly in many less-developed countries, systems of control over land and resources are under great stress, an important reality that deserves separate listing as an eighth and final opening observation. Sound local control would reflect both the traits and capacities of local nature and the needs and aspirations of the people who inhabit the local lands. Strong democracy need not be present; controls could take more hierarchical forms so long as adequate constraints are in place (cultural or social, for instance). In too many places, however, resource decisions are in effect made by outsiders seeking access to the local resources for their own gain, often short-term. They want access to oil, water, minerals, timber, rare species, or simply the topsoil and sunlight. And they are willing to cajole, bribe, or otherwise push hard to get that access with little regard for the wishes and needs of most local people.

Behind such moves is the reality of global markets and the ease with which resources extracted from one place can be sold for profit in another. Behind them too are simple scarcities of resources—not absolute scarcities, usually, but shortages in supplies that raises prices and heighten incentives to gain control over valuable resources wherever they happen to be. Thus we have today's ongoing "land grabs" and "water grabs" by sovereign investment funds, global corporations, and wealthy individuals, driven by profit motives rather than, typically, much regard for the plights of local people and the long-term ecological health of local lands. From the view of local people, the problem is mostly about external wealth that pushes them aside: The outside forces are to blame. From the view of outsiders, the problem can appear instead as local corruption and bad government generally, as a misuse of sovereign powers at the local level. In either story, however, local lands and resources end up losing in ecological terms. And so do the local people, who have been, in any fair assessment, dealt with unjustly.

The bottom line here, as on the last background point, is that systems of power and governance and the ills that afflict them have direct relevance to social justice and to the possibilities for good land use. Debate continues to swirl around what is termed the "natural-resources curse"—the too-frequent pattern that nations rich in valuable commodities (poorly developed ones, often) end up suffering for it, in terms of corruption, oppression, and warfare. But again, the causes of such ills need not be identified with certainty to conclude that social injustice is rampant and that it is strongly linked to misuses of nature. Good land use isn't likely to happen until the injustices diminish.

Normative Stances

With these observations set forth it is possible to identify the central normative stances that, in one form or another, seem necessary to bring about good land use and that are needed also to flesh out the full meaning of that overall normative vision.

Fair sharing. John Locke's liberal theorizing in the latter part of the seventeenth century largely began as a response to theorists then at work defending the claims of the Stuart monarchs of England. The monarchy's view was that the land of England in an important sense all belonged to the crown and that the private rights of all owners were based on the monarch's continued consent. In support, the theorists (Robert Filmer in particular) cited the passage in Genesis in which God gave the earth to the sons of Noah. These sons, Filmer claimed, took title to the earth not as individual owners but as putative kings of broad territories, and their claims to control descended over time, generation upon generation, to the current monarchs of the world. The Stuart kings thus controlled England through a chain of land title akin to the apostolic succession of the pope in Rome.

Locke's contrary claim was that the earth instead was given to all people in common, as a shared asset. Everyone was a co-owner. This beginning point got rid of the monarch's claim to land title, but it posed a new one that was, in fact, not at all easy to address. The new problem was that when an individual took control of a piece of the earth, converting it somehow into private property, the transaction involved, frankly, the theft of land from the commons. How could such theft be morally legitimate? Why would the other co-owners of the earth put up with it? Locke's answer, mentioned in chapter 4, was to draw upon the moral claim that a person who mixed labor with the land and thereby created value ought to own the value thus created. And when land was so plentiful that anyone who similarly wanted to create labor could just as easily do so, then one person's assertion of territorial control didn't limit the power of other individuals to go out and do the same. In a world of scarcity—Locke's world and even more our own—this argument made no real sense. But it was embraced anyway, because the bottom line—rooting land titles in some source separate from the crown—was the one people wanted to hear.

Locke's reasoning was hardly novel on any of the separate points in his argument. Each piece had been used before, including the idea that a person really didn't own a piece of land unless he had mixed his labor with it and created real value. (This meant, taken literally, that a person could not own unimproved land.) Also used before Locke was the story of a state of nature in which everyone shared the earth as co-owners. It was an alluring vision, one that seemed to make sense, not just to those who wanted to counter royal claims to control but to those who questioned (as Locke did not) the vast landholdings of the aristocrats. If everyone shared the earth together, with rights to enjoy its fruits as needed, then the vast landholdings of the aristocracy (and the church) seemed illegitimate. At the forefront of those questioning the agrarian order was the Swiss writer Jean-Jacques Rousseau, who blasted the arrangement in his critical inquiry on economic inequality:

The first person who, having enclosed a plot of land, took it upon himself to say *this is mine*, and found people simple enough to believe him, was the true founder of civil society. What crimes, wars, murders, what miseries and horrors would humankind have been spared, by someone who, pulling up the stakes or filling in the ditch, had cried out to his fellow men: "Beware not to listen to this imposter. You are lost if you forget that the fruits of the earth belong to all and the earth to no one."[2]

Rousseau's reasoning helped fuel the French uprising against the agrarian lords but few people then, as now, were willing to get rid of the private property system. What they sought was easier access to land by the landless and a more widespread and fair distribution of it. Thomas Jefferson was among those who agreed, even as he opposed any call to seize the lands of great landowners for forced redistribution. He proposed that government use every means possible to facilitate land-ownership (by which he meant small farms), or at least to provide comparable economic employment, thereby recognizing the moral claim of all to a share of the earth.

Whenever there is in any country, uncultivated lands and unemployed poor, it is clear that the laws of property have been so far extended as to violate natural right. The earth is given as common stock for many to labour and live on. If, for the encouragement of industry we allow it to be appropriated, we must take care that other employment be furnished to those excluded from the appropriation. If we do not the fundamental right to labour the earth returns to the unemployed.[3]

Jefferson's reasoning contained a distinct echo of earlier natural rights writing, particularly the principle that a person should not own and control more than he needed and could use. As Thomas Aquinas had put it, "Whatever a man has in superabundance is owed, of natural right, to the poor for their sustenance."[4] Particularly troublesome for Thomas More, as later for Jefferson, was the image of a large landowner who left lands idle. To the inhabitants of More's *Utopia* (1516), it was deemed "a most just cause of war when a people which does not use its soil but keeps it idle nevertheless forbids the use and possession of it to others who by the rule of nature ought to be maintained by it."[5]

As the eighteenth century progressed the idea of original co-ownership, the vision of a world in which everyone owned a share, transformed into a vague ideal of an individual right to property. That right was not chiefly a right to buy land on the market and hold it securely, although security of title was also talked about in these terms. Instead, it was an opportunity to gain access to land on easy terms—a realistic opportunity by laboring to become an independent owner. Jefferson sought to honor this claim in his home state of Virginia by providing in its state constitution that fifty acres of land would be given to "every person of free age" who neither owned nor had owned that much land. (Virginia did not adopt the idea; Georgia did.) This particular right to property drove efforts by the state and federal governments to make land readily available to settlers, often under one of the era's many homestead laws. In the mid-1790s, an elderly Thomas Paine weighed in on the issue, strongly agreeing that the earth began as "the common property of the human race" with each inhabitant "a joint life proprietor" of it. Private property arrangements, he recognized, made that co-ownership impossible in his day. But the moral right lived on. His proposed solution was to create a national fund "out of which there shall be paid to every person, when arrived at the age of twenty-one years, the sum of fifteen pounds sterling, as a compensation in part, for the loss of his or her natural inheritance, by the introduction of the system of landed property."[6]

These early views on the earth's co-ownership, and on the moral right of each person to share in the earth, are not easily or rightly pushed aside today. Defenders of concentrated wealth over the years would challenge this state-of-nature, co-ownership story, contending instead that the original condition was a world in which everything was unowned and that anyone who simply took possession of land (forget the need for any labor and value-added) gained rights superior to those of anyone else in the world. (In the United States the view was revived in a minor libertarian classic from 1985, *Takings*, by law professor Richard Epstein.) The nature-as-unowned view held obvious appeal to those content with economic inequality. But it did not supply anything like a morally satisfying answer to the questions posed early on by John Locke and then Rousseau: Why should others stand back and allow this to happen when their own liberties would be constrained by it, and why should later people, arriving on the scene, be content to have all of nature owned by those who came first?

Since these questions were first posed, human life has taken on ever greater moral value with a more widespread sense that all individuals, everywhere, are morally worthy creatures entitled to at least the basic elements to sustain life. Modern human-rights thinking, that is, aligns well with early natural law writing about the co-ownership of the earth. Details aside, the central theme has to do with sharing the earth so that everyone has a home and so that the needs of all can be met in fair ways. Sharing is plainly needed with respect to the good parts of the nature, particularly lands and waters that can supply food. Mutual respect would also seem to call for a sharing of the burdens that people impose by their living, sharing the ill effects, the pollution and contamination, to the extent it cannot reasonably be ended. First and foremost, environmental justice is about sharing the earth—its productivity and its capacity to absorb wastes—in fair ways.

Fair sharing certainly seems pragmatically necessary if people everywhere are to use nature in ways consistent with definitions of legitimate use. Poor people, as noted, will often do whatever is needed to survive, as will those who compete in markets that press producers hard to raise productivity and cut costs by misusing nature. This means ensuring that everyone's basic needs are met in ways that don't stimulate misuses of nature, and ensuring also that market forces are dampened enough so that mere survival is not conditioned on a willingness to degrade. It hardly needs saying that sharing on international scales is required to diminish wars and other conflicts over resources. Wars can stem from greed, of course, even (particularly?) by combatants who already have more than their shares. But war can also stem from a well-supported, anger-inducing belief that some people are simply taking too much.

Beyond that, good land use would seem to require communities of land users and land dwellers whose attachment to a place, and whose support for local laws and ownership arrangements, is such that they feel bonded to the place over the long term and are invested in its flourishing. The allocation of land-use rights must make normative sense to them, even if unequal. They must have long-term security in their patterns of use, preferably with a confidence that lands can be passed along to later generations. And they must support the legal and cultural schemes that constrain land and resources uses so as to keep landscapes healthy, setting expectations if not coercive demands that people act right. Support for such rules is not likely to come about unless the overall system seems fair. (This lesson has come out very clearly from recent scholarly investigations of common property regimes and the elements that must be in place in a regime, and among a land-use population, for the collective control of the common property to work well over the generations.)

What then might a fair-share arrangement look like? How would it operate at various spatial scales? And how would it be put into effect when we take into account (as we must) that some lands are much more hospitable and productive than others?

The questions are hard ones. They are also—more pertinently— morally inescapable ones; they are questions that, once answered (tentatively, given that no answers are likely to be final), provide key elements of visions of good land use. Good land use means people living in place in settings in which they accept land allocations as reasonably fair, in which they support the rules and norms limiting who can do what to nature, and in which they feel committed for the long run. The basic needs of all should be met to avoid both the social and political stresses that poverty can engender and to forestall the kinds of desperate survival means that degrade so many landscapes.

This vision of sharing will likely require for its achievement rather significant limits on the role of the market in allocating resource-use rights and on the means market participants can use when seeking profits. The market alone, with government merely standing by to protect private rights and keep the peace, cannot be relied upon to generate anything like fair-share allocations. Not even close. They also cannot be left to operate freely with any real expectation that market competitors will respect the autonomy and long-term security of local land users.

The market considered as an allocation method is usefully contrasted with the model of the family or tribal group, or any setting in which everyone is included and everyone's basic needs are met. The latter is a setting in which shifting needs lead to shifting mixes of use-rights. In many tribal settings, tribal elders often made decisions about who could use what and for how long, but they did so, and here and there still do so today, within moral contexts in which their duty is to ensure that the needs of all are met. Typically, nothing like full economic equality would seem needed. But a system must seem fair to the people covered by it; it must receive their assent and cooperation.

Given how people today are pressing beyond the planet's carrying capacity, and given how rising national incomes go disproportionately to the already wealthy, it should be clear that the fair-sharing component of good land use calls for an individual ethic of consumption. This, too, would seem a necessary part of an overall vision of good land use. The wealthy must constrain their appetites, particularly in terms of their consumption of resources that come from the nature. Our overall ecological footprint needs to get smaller. And this needs to happen even as the planet's poor gain larger shares of the planet's produce. Increased technological efficiency can of course help, as can resource reuse and recycling. But the wealthy also simply need to use and consume less physical stuff.

Local attachment and control. Individuals engage in good land use, it seems, more often when they are embedded in local communities of land users that are attached to the land, have long familiarity with it, and see good land use as a duty owed to the multigenerational community as such. They need to be embedded, that is, in a social order that brings out the best in them. They must think of their local lands as places to live, not opportunities for exploitation. They need to be connected to the fund of inherited knowledge on how to use local lands well, which is to say part of a network of knowledge sharing that reflects a sense of community and membership.

Good land use will often call for the coordination of land uses at fairly large spatial scales. Longstanding irrigation arrangements often reflect this kind of coordination. So do land-use planning schemes. No individual landowner along a river has the power to take good care of the river; that task must be shared by all. No landowner acting separately can provide room to meet the needs of wildlife, particularly species that migrate. The protection of coastal barrier islands and floodplains requires organized effort. Large-scale good land use, in short, is not possible for individuals acting independently; it requires orchestrated effort. This calls, in turn, for a social order and a regime of laws or other norms that facilitates such organized effort and that encourages individuals to engage in it. It calls also for sufficient autonomy from external interferences, including market pressures, that frustrate this kind of organized collective conservation.

Autonomy for local land users does not mean that outsiders all need to stay away. Instead it means that they should respect local decision-making and not disrupt it. It also means that they should avoid activities where they live that would have spillover effects interfering with good land use elsewhere—pollution that carries, or inappropriate interferences with waterways. Local people need the freedom and capacity to practice the art of good land use and outsiders need to allow them to do so. That is the guiding light.

This component of good land use will likely require significant limits on the market's operation—limits, for instance, on absentee ownership of land and resources and limits on the ability of outsiders with money to distort local decision-making. It will also require rather significant cultural shifts. It will require a considerable change in the idea that those with money can buy what they like, and that if they have purchased an item fairly in the market it is rightly theirs to own and use. There are many reasons why this normative claim, while acceptable in many settings, needs significant pruning. Limits already do exist, on the purchase, for instance, of products from endangered species. Here and there in the American West, rules governing water allocations limit or prohibit outsiders from entering a catchment basin or coming to an aquifer to extract water to ship and use far away. Nations have long had laws limiting land ownership by foreigners. Far more of these laws are likely to be needed to support and protect the independence and cohesiveness of local land-using communities.

More generally, respect for local control and autonomy will require changes in basic notions of just deserts and in accepted notions of individual liberty. Such changes go to the very roots of modern liberal culture, particularly as it has been refined and strengthened in recent decades by market-based thinking and neoclassical economics. It calls for changes to the notion that a person deserves whatever she can get through market transactions, with no concern, for instance, with the ways that other people, present and past, have helped make the production or acquisition possible. It calls for new understandings of the idea of harm, as it is used in the general liberal (and land-use) principle that liberty exists only so long as one does not cause harm to others or to the community as such. Perhaps above all, it calls for changes in views of nature itself, particularly to the idea that nature is best understood as a warehouse of commodities awaiting human use. The better view, the ecological view, is that parts of nature everywhere are embedded in ecological systems and play roles in sustaining the functioning and productivity of such systems. In the ecological view, far more attention is paid to the line between principal and income, between using the land's produce—the part that can be safely taken away—and cutting into the ecological principal.

Expanded responsibility. A third main component of environmental justice can be covered more quickly. It is a normative application of the basic understanding of interconnection, a normative recognition that no action in the physical world takes place in isolation but, to the contrary, every action is connected by long chains to much else that occurs.

A common way for contemporary people to separate themselves morally from ecological decline is to assume that the market somehow cleanses both physical goods and physical wastes. The reasoning, so to speak, is familiar. When an item is purchased and the price paid, the buyer has no moral connection to the product's history, to how it was made and shipped, by whom, at what costs. Similarly, when waste is turned over to a waste hauler, the generator of the waste has no moral connection to its disposal and thus no connection to the lands and people who are affected by the waste. The out-of-sight-out-of-mind phenomenon explains this in part. Also at work is the sense that, by acting this way, a consumer respects that autonomy and decision-making power of other market participants. The market cleanses, that's the basic idea. Goods arrive on the shelf not just bright and cleanly packaged but cleansed of all ills or evils in the production and distribution process. Who wants to hear about how the thundering low vibrations generated by massive ocean cargo ships can disrupt the communications of sea mammals? Who wants to hear how leaching chemicals used in mineral extraction can leave long sections of rivers dead of all fish? Who wants to ponder the images of mountaintop removal mining?

The issue here is not a simple one. It is hardly reasonable to expect consumers to know the history of every item they purchase. But moral responsibility need not just take the form of changed individual purchasing decisions; indeed, that role should be secondary. Information-processing tasks that are overwhelming for the individual are less so when people work together. Product-certification organizations can help clarify which products have better histories to them (recognizing, of course, that "better" requires use of a normative standard on which fair minds can differ). Even more, governments can and should take more active roles in screening products to press against actors and practices that are clearly damaging. Individuals can show their acceptance of responsibility by supporting such legal rules and calling for more and better ones—that is their main, much-needed role. Much the same can be said in the case of waste handling and disposal.

The central reason why the market does not cleanse is that markets are competitive and participants in them are pressed to cut costs. It is the way that markets work, not an incidental effect. Cutting costs can very often include environmental degradation (and other ills, including labor abuse). To allow competitors to degrade as a way to cut costs is not to respect their decision-making independence. Nor is it to respect the sovereign powers of the governments where producers live to make their own tradeoffs between profits and healthy lands, though there is something to be said for that view. Some production practices are simply unacceptable as such, the ecological equivalents of using slave labor. And it is not enough to leave it to a community whose economic survival is based on misusing nature to stop doing so. Misuse is built into the system, and market participants are often bound by it (or reasonably see the world that way). For the abuse to end the system needs to change, plain and simple. Support for change needs to be more widespread, coming from those who, as consumers or other product users, gain from the misuse. And the support needs to take the form of calls for new laws and public policies.

The tendency to defer to local sovereignty—to claim that local peoples should make their own tradeoffs between jobs and degradation—is even less defensible when the local people are driven by poverty and when decisions are made for them by corrupt officials or institutions. Poverty is itself often a byproduct of global markets and bad government. Bad government, in turn, while often having many causes, is rarely separate from global markets—particularly the natural-resource curse. It is rarely separate from strong outside pressures, above the table as well as below and from international organizations as well as corporations. A third-world country that allows a global corporation to clear-cut a rainforest, long used lightly by locals, is not likely making a decision that reflects anything like true local sovereignty. It is disingenuous to claim otherwise. Many governments seemed locked into continued degradation by debt payments, apart from any bribery. Nor is the problem limited to international contexts. Farmers in the American Midwest are the region's predominant water polluters, in some states (Illinois) accounting for much more than half of all water pollution. Farmers use polluting practices because they feel that they must (often to compete for leases requiring high cash rents); the state in turn does little because the political power of agribusiness is so great. Disinterested citizens, supposedly the holders of sovereign power, play essentially no role in the decision-making.

Just as the comparatively wealthy need to embrace an ethic of consumption, so too all consumers need to accept their moral connection to the goods and services they enjoy and the wastes they generate. Only in part does the market diminish that connection. Only within limits is it fair to deny responsibility by pointing to the free choices made by other sovereign actors.

Migrants and making room. Climate change, as newspapers now point out, will force many people around the world to leave their homes and move because their homes are no longer habitable. Such migrants, though, could make up only a portion of those who need to move due to climate change and other ecological ills.

Many rivers today are unhealthy and can regain health only by altering how they are managed physically (for instance, by breaching levees that keep water away from natural floodplains) and by the locations and types of land uses in their catchment basins. Many wetlands, barrier islands, and other coastal zones should not be inhabited or intensively used by humans. Those living on them need to change their locations. Wildlife corridors need constructing and protecting, which means more people being asked to relocate. Much irrigated farming needs to end because water supplies are insufficient and the irrigation pollutes waters and degrades soils. They too may need to move.

When market forces compel people to move, as their jobs end or businesses dry up, we often look the other way, counting this as a cost of competition that is presumably beneficial overall. The rationale can be questioned; it certainly can be said that calculations of competitive efficiency ought to start taking these massive costs into account. Even less justification for denial, though, is present in the case of migrations that happen either due to environmental degradation (rising sea levels or salt-water intrusion, for instance) or due to organized efforts to change land uses at large spatial scales—for instance, moving people out of floodplains to allow the plains to perform their longtime ecological functions. Such changes need to take place for the good of all, in many settings the sooner the better. But they will be hard to bring about without a shared recognition that, because we have all helped bring on the problems, we should all, in fairness, pitch in to help solve them.

The implications of this simple conclusion are many. The chief cultural ones have to do with the ways we think about boundaries and territorial rights, both those based on private property and those on sovereign power. In the United States, popular sentiment needs to shift far away from its current, ill-considered hostility to the use of eminent domain (expropriation) powers to force people to relocate upon payment of fair compensation for their losses. Today's current hostility toward expropriation—far different from the prevailing attitude when interstate highways were being built—is emblematic of many of the cultural elements most in need of change, above all an exaggerated individualism and denial of community. Even more, there is a need simply to make room for people, particularly people who will legitimately desire to move en masse. And there are the obvious costs involved in this, and the need to share the costs equitably.

Entrenched notions of private ownership and individual entitlement will frustrate such measures and need to be challenged head on. Even tougher will be the tendency of people pretty much everywhere to take their sovereign boundaries too seriously and not recognize that, while appropriate and often necessary, the boundaries need softening (that is, to become more permeable) given the moral claims that people have to share the earth. Those forced to leave their homes, by natural causes or organized planning, need to find new homes, which means people elsewhere need to open their borders for them. This is particularly true in the instance of wealthy countries, which are disproportionately responsible for global ecological degradation and better able economically to help.

The Responsible Land Owner

One of the central shortcomings of the environmental reform move-ment in the United States has been its failure to give real thought to the institution of private property rights in nature and to craft new ways of thinking about it. Scratch the surface of pretty much any land use–related environmental dispute and one almost immediately encounters resistance to remedial measures grounded in claims that the measures would violate private rights. As resistance has grown, particularly to measures at the federal level involving wetlands and endangered spe-cies habitat, starkly libertarian visions of private ownership have risen up and gained currency. As regulation opponents have waved the flag of property rights, the environmental side has largely remained silent except to contend that regulations are worth their cost.

What has been missing here, sorely, is a thoughtful, well-crafted vision of responsible land ownership in which it is sensible to expect landowners to act in ways consistent with the health of the land, in which the rights of owners to use nature are tailored to take into ac-count the peculiar features and ecological context of the nature that they own. The environmental movement should itself stand in favor of private property, and not get painted as its opponent. It should step up to frame the issue: The environmental side is in favor of respon-sible landownership; its opponents are out to support irresponsible ownership.

An effort to reform private property needs to begin with the recog-nition that property ownership exists in two forms. It is, technically, a legal institution with rights prescribed by law. Jeremy Bentham put it aptly two centuries ago, "Property and law are born together and die together. Before laws were made there was no property; take away laws and property ceases." Yet property, particularly in the United States, also has a cultural life to it. It exists in the public imagination and is linked to the nation's self-image as a land of liberty and opportunity. People are quick to proclaim what ownership means when they would hesitate to offer an opinion on any comparably complex body of law. This cultural embrace is linked to the common view that property is somehow an individual right that exists apart from government and laws, something of independent origin like the right of free speech. This isn't true; the constitution (as the Supreme Court has said many times) only protects, as private property, rights that arise under some other, nonconstitutional law. But popular images of ownership carry great weight. A reform effort thus needs to aim at property as a cultural creation, not just at the legal reform.

Property reform is an environmental justice issue given the ways that control over nature brings control over other people, limiting their ability to gain access to nature and share in it. Also, property regimes deemed unjust can undercut the kind of community collegiality and long-term commitments to place that seem essential for good land use to take place.

Property reform, though, is also of critical importance for reasons that go beyond social justice. Understood as an institution, private ownership is one of the chief ways that managerial powers over nature are delegated to particular individuals, who, given the realities of interconnection, make land- and resource-use decisions that affect other people as well. Good land use might well not be possible without owners who care about the land and are bonded to it over the long term. But not every owner will fit this description, and even those who do will face strong pressures to enhance short-term production at the expense of lasting health. Even well-intentioned landowners may not be aware of the harms they cause or be able economically to avoid them given market competition. Then there are the land-use changes that need to be orchestrated at larger spatial scales and are possible only if governing powers can require landowners to act in particular ways.

The subject of reforming property law is a large one, and much can be (and has been) said on it. Here, four main points might be made:

First, owners of lands and other parts of nature need to limit their uses of them to those that are ecologically sound, in the sense of consistent with an overall vision of legitimate land use. Owners need to be asking, not just "what can I do here to make the most money?" but also "what can I do here safely given the ecological features and attributes of my land?" Property rights cannot be defined in the abstract, as law theorists have often done—writing about the rights of ownership in a hypothetical Greenacre that lacks any ecological features or context. Instead, ownership rights need to be tailored locally to the peculiarities of place so that owners confine their activities to those consistent with the good use of the overall landscape.

Second, one of the main ways that this tailoring of rights can take place is to tinker with the longstanding idea of land-use harm, the principle that, while owners can use their lands (and waters and more), they must do so only in ways that avoid causing harm. Harm can be to neighbors; it can be the surrounding communities or to people downwind or downstream; it can also be, it needs to be, to future generations who will use the land and, in some way, to the other life forms that also call the land their home. This particular reform (redefinition of harm) could be easier than others because the do-no-harm rule is familiar and entrenched. The challenge is to put forth a new definition of harm that gains acceptance.

A knotty part of this challenge will come from those types of land-use activities that are harmful not in isolation but only when too many landowners engage in them. As examples we can cite catchment basins with too much permanent cover removed or too much land drained or too many chemicals allowed to run off. We might call these carrying capacity harms because they involve landowners collectively going beyond the land's capacity to support a particular type of activity. These harms will be harder for people to see because they often involve actions that seem harmless enough in isolation, and might well be—actions that would be perfectly fine if only a few people inhabited the landscape. In such settings, the key rhetorical claim might be best based on a call for people to do their fair share or their bit (a moral claim that resonates), for people to pitch in and do what is needed to modify their activities to avoid the harm.

Third, it is essential today, as many people have seen in the past, that key parts of nature be retained in public ownership and managed more directly by collective means in the public interest. Under long-standing legal doctrines, water, wildlife, navigable rivers, and beaches have all, in the United States, been commonly viewed as public assets, owned directly by the people with the state playing the role of trustee and manager. Such components of nature, *in situ*, have been understood as impressed with a kind of public trust so that decisions made about them are reviewed closely to ensure they do not materially interfere with the special roles that such resources play. These key parts of nature need to remain in that category, with governments at all levels bound to protect them from improper use or degradation.

As prominent scholars have pointed out, particularly law professor Mary Wood, the rationales supporting this public-trust status for these key resources could apply just as sensibly to other parts of nature. And the need to apply it is now great. For many ecologists the most vital of all natural resources is topsoil, the fount of nearly all terrestrial productivity and life. It simply must be kept fertile and in place, not degraded or eroded. Even more pressing is the atmosphere and our need to protect it from degradation, climate-changing gases above all. Mary Wood has spearheaded an effort, brought on behalf of children, to press governments in the United States to do more to protect this shared resource by curtailing damaging emissions.

Two Confusions

Two further topics need to be touched upon before drawing the material in this chapter to a close. Talk about environmental justice today, particularly at the global level, tends to focus most often on climate change and the limited ability of the atmosphere and oceans to absorb greenhouse gases. As noted, there are good reasons why the atmosphere (and oceans) might properly become special, public trust assets and no longer treated as unowned. The climate-change problem also illustrates well the reality of ecological interdependence and how people in one place need to accept moral responsibility for the effects of their lifestyles on people elsewhere. The central moral issue at stake—aside simply from the need to draw a sensible line between use and abuse—has to do with how we share the atmosphere's carrying capacity once we have decided what that capacity is.

A common stance in the United States—indeed, so common that almost nothing else is heard publicly—is that international agreements to combat climate change should start with today's emissions levels and require everyone around the world to cut back by the same percentage. This approach, of course, ignores the huge variations in emission levels today. It also ignores emissions levels in the past, which were even more concentrated in the hands of a small number of polluting nations. Finally, it ignores the needs of people around the world to grow their economies, perhaps following trajectories similar to those followed by today's developed nations.

This approach—using today's emissions levels as the point of beginning—has hardly any merit to it in any moral sense, not after one denies the moral relevance of being first-in-time to pollute. A more plausible beginning point is that the carrying capacity of the atmosphere should be, as Locke, Paine, and others argued, the common property of all with each person enjoying an equal right to use it. This moral stance would seem to point to an allocation of pollution rights based strictly on population. Countries such as the United States would need to make massive cuts in emissions just to get down to their proportionate share of current emission and before even dealing with the collective need to curtail those significantly. Going further, of course, a moral inquiry could quite fairly consider histories of emissions and charge those emissions to the nations that made them. Using that calculation, an industrialized country such as the United States, Great Britain, or Germany would need to reduce its future emissions even more drastically, given how much they have polluted already.

This latter approach is infeasible. Beyond that, good arguments can be made that the approach is unfair to those living today in developed countries who were not personally involved in polluting the atmosphere generations ago. Still, the dominant stance of the United States is equally flimsy in moral terms, and needs to be understood that way by Americans.

Those living lives of high carbon emissions are causing harm to people all around the world. Under a sensible liberal standard of do-no-harm, the behavior is wrongful. That said, the problem is a shared one and solutions to it call very largely for collective action, particularly in the ways energy is generated and supplied. What is most needed are not changes in individual behaviors, though changes are in order. More important is the need for people collectively to support, and insist upon, immediate changes in laws, policies, and institutions.

As for the second topic, a bit more needs to be said about the paradigm case of environmental justice as the term is often used in the United States, the case of a hazardous landfill or other polluting or unwanted land use that is sited in a location where it disproportionately harms minorities (or, in some versions, poor people). The charge is basically one of racism—"environmental racism" it is called.

Racism is inherently bad, and we can criticize it as such. But to what extent is this paradigm case simply a problem of racism? What if the landfill were sited instead in a neighborhood where the racial composition of the local residents matched some regional or national average so that impacts were not disproportionate. The racism would end but what about the environmental problem? Would moving the landfill take care of the concern?

The answer is plain. The local people would still have experienced an unwanted land use brought into their neighborhood and causing or threatening harm. The *social* harm would be different and less with the racism gone. But the *environmental* harm would be just the same. Just as many people would be threatened. Just as much nature could be degraded. To see this is to reveal why a focus on the racism aspect of this scene comes at a real cost because it diverts attention from other aspects of the scene that are also troubling.

The environmental component of the problem plainly includes the hazards and harms imposed by the landfill itself, on surrounding nature and the local people, without regard for race. The problem also has to do with local control and with the ability of local people collectively to insist—of one another and outsiders—that their lands be used in ways that qualify as good land use. The story has imperialistic overtones (although, of course, important details are lacking) and overtones too of using nature as a tool of exploitation. The story calls into question the moral legitimacy of a private property system that allows lands to be used this way, and raises questions about the power and proper realm of market forces. Have those who generated the wastes being disposed shown adequate recognition of their moral connection to them, and what about the impacts on public trust assets?

These distinctly environmental concerns are all part of this story. A racial focus, while clearly important, can make them harder to see.

Virtue and Community Membership

In his useful survey of approaches to justice, Michael Sandel observes that older approaches to justice—those in the world of ancient Greece and Rome and, in revised form, in the medieval West—were largely based on personal character or virtue. As Aristotle expressed it, we should strive for moral excellence in what we do, strive to live honorably. The chief modern approaches to justice, in contrast, begin with the ideal of individual freedom and seek to enhance it for all while recognizing that acts by one person can interfere with the liberties of others. Given this social interconnection, individual liberty cannot be absolute. In one modern version (as we have seen) we should look to the consequences of our actions (their effects on utility, of ourselves and others) and avoid conduct that generates more harm than good. In the other popular version, also noted, we are all rights-bearing individuals and should act in ways that respect one another's rights. Of course many writers on justice mix these categories or go beyond them, but they are accurate enough to highlight how ideas of justice reflect the main elements of modern Western culture with its emphasis on individualism, human exceptionalism, liberty, and autonomy. To the extent (if any) that moral norms have objective existence apart from social convention, they simply reflect these now-dominant cultural components, vesting individual humans (and no other life forms) with rights.

On the topic of justice a reformed culture, adequate to promote good land use, needs to take a much different shape, perhaps looking back more (as Sandel encourages us to do) to the older, virtue-based approaches to justice, but perhaps breaking more new ground.

Social justice thinking, like our thinking about nature, needs to begin with the land as an interconnected, interdependent community of life. It needs to present the individual human not just as an autonomous being (an element that does need retention, in diminished form) but also as a member of the land community and a member too of the social order. We are members of various communities, starting with the family and expanding outward. Our social justice—as philosopher Baird Callicott explains—needs to be based on these community roles. It needs to be based on what he terms "sentimental communitarianism," which is to say our fact-informed, genetically influenced, reason-tailored senses about our communities and how we can uphold them. This approach bears resemblance to older, virtue-based approaches. But such approaches can be (as they have been) presented so that they too treat humans as individual beings, as the sole holders of moral values, who begin from positions of autonomy. Virtue-based approaches, that is, can have the same flaws as other theories of justice.

It is in the context of our various, overlapping communities that we need to hammer out anew our senses of right and wrong living with one another. This work needs to begin, as repeatedly urged, with the land community and normative stances on its health and good use. This includes some expanded form of the idea that key parts of nature—perhaps most or even all of it—are best understood as the common property of all, as public trust assets, which individuals can use so long as the long-term public interest is respected. Fair-sharing of nature needs to be just as central, a normative principle with many implications both for individuals (an ethic of consumption and call to collective action) and for political entities (a willingness to make borders permeable and to accept collective responsibility for past acts). All private rights in nature need restructuring to make them morally legitimate, so that they do not serve on balance as tools of domination and exploitation or empower owners to degrade the nature they control. Such rights need to be embedded in systems of community control that appear fair to local people, that are free of improper outside interference, and that successfully invite people in their land uses to rise to their ethical ideals. High hopes, to be sure.

It should hardly need saying that this approach bears only the most superficial resemblance to the concept of sustainable development pushed in the international arena. Like other versions of sustainability the term, as a normative standard, is vague enough to provide a big tent for all manner of interests, including those bent largely on pushing the range of global capitalism with little regard for environmental consequences. In the quite different normative vision sketched here, good land use remains at the top of the normative list and all else is judged by it. There is no muddying the goal by mixing it with desires for expanded market production. In its common form sustainable development (as historian Donald Worster observed not long after the term gained currency) suffers from too much inner tension; it is a pair of horses, supposedly attached, that want to run in different directions. As such, the goal provides for too-easy ways to justify continued degradation. It misses most of the social justice issues, including those having to do with local attachment and control and morally sound private property rights. Most of all, it aims to address poverty by growing the global pie, which is to say, inevitably, by overtapping nature even more.

Good land use needs to remain a separate goal and not be mixed with anything else, even as efforts to achieve it necessarily require substantial progress also in dealing with other goals. Such other goals can and should exist, of course. And tradeoffs among them will be necessary. But tradeoffs need to be done overtly and recognized as such. They cannot be buried in the details of a sustainable development report that claims to be making progress overall even as ecological degradation worsens.

SEVEN

The Capitalist Market

In the intellectual effort to escape Plato's "tyranny of the present" few challenges rate higher than making sense of the capitalist market and its potent messages, seeing how it reflects and, in turn, shapes and strengthens particular ways of perceiving the world and valuing it. A critical look at market capitalism is also essential to trace land degradation to its root causes and identify possible reforms. Many market critics over the years have pointed out the various ways market incentives can encourage misuses of nature, even as the market can also invite owners to keep lands productive. This intellectual terrain needs a quick review. Less-often explored are the ways markets and market-thinking frame our views of the world. They affect how we see ourselves as well as nature and how we go about judging behavior, our own and others. They also affect our inclinations to work together collectively. These days, national well-being is measured chiefly in terms of the volume of market transactions; we are as good as our market is large, so we assume. In the world we know, the market looms large.

The inquiry in this chapter takes a probing look at capitalism, defined here in its classic sense as a system in which the owners of productive assets (capital) use them to generate profits that are then plowed back into further capital assets so as to generate greater future profits. It also looks at the market, understood as the social realm in which people buy and sell goods and services, including rights to use nature. Capitalism can exist (and has) in settings where markets play relatively small roles, settings where production and allocation decisions are mostly made bureaucratically. (The former Soviet Union's system, perhaps best described as bureaucratic state capitalism, illustrates this possibility.) It is market capitalism, though, that dominates in the United States and to varying degrees elsewhere, even as economies everywhere include sizeable components that employ other allocation means (for instance, public schools, police and fire services, highway building, much health care, and many utilities). Finally, the chapter pulls in strands of economic reasoning that justify market capitalism and its underlying private property system. It is the combination of these elements—capitalism, market allocation, economic reasoning, and private ownership—that frames so much of our worldview, that exerts such influence on how we perform our oldest task.

As we shall see, the market as a venue for exchanging goods and services is not itself the major impediment to good land use. Nor is private property, per se, a problem; indeed, a well-designed system of private ownership can bring many gains. The problems lurk just beneath the surface, in the constellations of values, understandings, incentives and judgments that undergird the capitalist market. Put simply, the capitalist market of today, particularly as it rises above state power, is the embodiment and citadel of most of what is flawed about and within modern culture. It is flawed as a scheme of values, as a lens for seeing nature, and, freed of appropriate limits, as a way of linking people together and organizing human action.

Perceptions and Values

Systems of private property (and their more state-run alternatives) begin their operations by dividing nature into tracts of land and into other, legally defined rights to use nature (water rights, for instance). Management authority over these divided pieces is then partially turned over to owners, private or public. The delegation of power is only partial because rights to use pieces of nature are typically limited by law, sometimes considerably. Also, governments reserve power to regulate property uses over time and otherwise to refine the exact powers that owners wield. As the market goes about its business, these pieces of nature are bought and sold and variously used, consumed or saved, all as their owners see fit. Items that trade in the market take on exchange or market values, sometimes based on the free play of pressures to buy and sell, sometimes affected by the powers of one side or the other to manipulate the trading price. All of this, of course, is familiar, so much so that the arrangement's chief elements hardly draw much comment.

Not really caused by this arrangement, but certainly incorporated into it, are a number of critical, troubling ideas about nature and its values. As often noted, market capitalism has the effect of dividing and fragmenting the natural order, which is, ecologically speaking, highly integrated. The resulting pieces and parts are what individuals and entities can own and what they buy and sell. These are the things that take on market prices and get assessed in terms of price. The value of a particular thing—a tract of land, for instance—is certainly not unrelated to its surroundings; a lakefront home, for instance, is worth more because of the adjacent lake. As realtors say, value is all about location, location, location, which is to say proximity to other land uses and amenities that make a given tract more or less valuable. Nonetheless, parts of nature are mostly thought about as discrete parcels or resources with particular dimensions and physical features. They are priced and traded that way.

Guided by this market approach, it becomes easy and natural for people to think of nature itself in such fragmented terms. Value is not placed on larger community-level or landscape wholes because they are not marketable as integrated wholes. Similarly, parts of a landscape are not valued in terms of their functional contributions to the whole. With market value so influential it becomes harder to think of valuation set in other ways; wetlands, for instance, are not thought about as functional systems that help remove silt and pollution and as nurseries for aquatic life.

Buyers for the most part value a part of nature based on the benefits they obtain from owning and using it (and ultimately selling to others for ownership and use). The benefits that get considered are only those benefits the owner can capture personally. Thus, the question for a buyer is not: What benefits can this piece of nature produce overall? Instead it is: What benefits can I get from this piece of nature, now or very soon, ignoring benefits that run to other people elsewhere and in the future? Benefits that go to other people—neighbors, most often, but sometimes people farther away—are easily ignored because the other people are unlikely to pay for them. On the flip side, the external harms that a land- or resource-use imposes on neighbors can also be ignored in calculating asset value when, as often, the owner generating the harms doesn't have to compensate people injured. To use the common terminology, asset valuation ignores land-use externalities, both good and bad.

This familiar, market-based way of seeing and valuing nature contains and accentuates particular messages, ones that have been flagged and challenged already:

· For starters, nature as thus fragmented is thought about as a collection of tracts and resources, available for purchase like so many items on a grocery shelf. Some parts of nature are valuable; many others are essentially worthless.
· In moral terms, humans living today (those with money to spend) are the actors. Nature's parts are mere objects; morally empty commodities that gain value only insofar as people spend money for them.
· Humans are the conquerors and owners; nature is the stuff that is controlled and consumed.
· Nature as an interconnected whole is not really noticed nor are its connections, processes, and interdependencies.
· The quality of the human environment, in terms of aesthetics, healthfulness, and convenience, also mostly goes unnoted and unvalued in the market except insofar as it affects the market prices of parts and as owners of parts see money in enhancing such amenities.
· Similarly, future generations of humans are valueless and absent because they are neither commodities themselves nor actors with money to dispense.

In this market-based view of the world, the economy is central to all that goes on. Everything else—nature and people—is attached to it and largely defined by it. People are categorized by their roles as producers (sellers of labor, mostly) and commodity consumers. The present is valued more than the future with the result that future costs and benefits play smaller roles in the valuation process, roles that can diminish over time (varying with the discount rate used) sufficiently fast that they become irrelevant when looking ahead fifty or eighty years. The market, in other words, encourages if it does not demand a short-term attitude toward valuation.

At the aggregate level, the health of a market economy is typically measured in terms of market activity; in terms, that is of the aggregate market prices (or similar measures) of goods and services that change hands during a defined period of time. Overall summations of market transactions (GDP) pay little attention to the quality and types of goods and services being traded; all of them enter into the summation, whether or not they are beneficial under any sensible normative evaluation. Externalities are similarly ignored in such calculations, including environmental harms, as is the exhaustion of the parts of nature that owners use.

As for the social order, market-based thinking portrays people mostly as autonomous individuals, as buyers and sellers of labor and goods. In capitalist legal systems, owners of wealth typically find it easy to pool their resources and to act collectively (as partnerships or corporations, for instance). Workers, in contrast, encounter a good deal of trouble trying to pool their labor resources and to act in concert. The difficulties that labor faces in the United States (more than elsewhere) have much to do with the sustained hostility by big business to labor unions, reflected in pro-business labor laws. Tellingly, anti-union rhetoric highlights the importance of individual autonomy and the primacy of (negative, individual) liberty, which are supposedly weakened by unions. Anti-union rhetoric, that is, further accentuates the market's tendency to portray people as detached individuals, not in terms of their various roles in social and ecological communities. Further, it illustrates the market's tendency to encourage people to act on their personal preferences, promoting individual welfare rather than any vision of collective good.

Cast aside in this market-based view of the world, as already noted, is any sense of a person as a citizen, one who gives thought to the overall welfare and who pursues it through civic means. Similarly, no thought is given to a person as a member of a land community except insofar as that role shows up in individual decisions to buy and sell. As for human communities or societies, they exist as collections of individuals who can associate with one another, or not, as they please. People are not members of communities (social or natural) against their will. They are not defined in important part (ontologically) by any unchosen interconnections. Future generations, again, do not show up as moral actors nor do those who are dead and gone. They show up, much as other life forms show up, only insofar as they are reflected in the decisions of today's market participants.

Normative Judgments and Responsibility

The above comments on the market's valuation of nature connect to the larger topic of how the market influences our capacities to engage in normative judging. The topic includes not just thoughts about the common good and, pertinent here, how best to evaluate land uses. It also extends to our capacities to evaluate actions, both our own actions and those of others. When and how do we judge behavior, and when and why do we take responsibility for the consequences we bring about?

Perhaps the most common moral complaint about the market, raised by observers from across the political spectrum, is that it puts a high stress on acquiring things and on ever-expanding wants. It breeds dissatisfaction with what we have and makes us yearn for newer and more. Wendell Berry summarizes the complaint:

The scarcity of satisfaction makes of our many commodities, in fact, an infinite series of commodities, the new commodities invariably promising greater satisfaction than the older ones. And so we can say that the industrial economy's most marked commodity is satisfaction, and that this commodity, which is repeatedly promised, bought, and paid for, is never delivered.[1]

Dissatisfaction is, of course, intentionally bred by advertisers who implicitly discredit the old while hawking the new.

Along with this commonplace lament is the longstanding concern that the capitalist market with its competitive individualism tends to undercut the very ethical values of trust, honor, and professionalism that are essential for the market's own efficient operation. This concern was prominently expressed a century ago by Max Weber and later reformulated by many others, including Karl Polanyi, Richard Weaver, and Daniel Bell. As noted, Adam Smith in his apology for a free market presumed that participants would remain guided by Christian ethics, just as John Locke had done in his liberal writings in the previous century and John Stuart Mill would do (less confidently) in the nineteenth century. This complaint rings true today, embedded in headline news on deceptions by financial companies. It is usefully expanded to take into account the strong, often successful exertions by major market players to distort public lawmaking, to halt needed regulations, and to inhibit law enforcement. Among the most amoral market players—and the market often rewards moral laxity—law and morality are rules of conduct easily violated so long as the chances of detection are sufficiently low or when the penalties, discounted by the chance of nonenforcement, can be absorbed as costs of doing business.

The market, as noted, is distinctly present-oriented in ways that discount the future and detract from worries about it. Only people living today are morally worthy actors (to the extent of the money they have to spend); all others have market value only insofar as a market actor takes them into account. The same precarious moral status attaches to other life forms, which also enter market thinking only as commodities or insofar as market actors otherwise pay attention to them. The market, to be sure, does not preclude recognition of moral value in future generations, other life forms, and people who are penniless. It simply makes such value optional with market participants acting on their personal preferences and the value lasts only so long as the preferences do. The market at its best gives people what they want at low cost, without passing judgment on the goodness or badness of such wants except insofar as laws adopted through nonmarket means constrict choice and thus the market's operation. When the market is put in the center, normative choices are largely left to individuals and public discussions deal with facts and reason: The market thus accentuates the cult of objectivity. Its much-touted virtue is efficiency, which is, of course, not an end or goal but simply a trait of the means used to achieve a goal; a desirable trait in some settings but not others.

The market largely accepts the normative choices of individuals with money to spend. It is equally accepting of goods offered for sale with little concern for who made them and at what environmental cost. In doing so it implicitly validates these normative choices and production processes. In this function it works in tandem with the cultural version of private ownership, which treats actions by owners as their personal business so long as no harms are imposed on others. People with money are free to spend it as they like and the preferences that guide their spending are simply their own business. Similarly, those who use land are free to do so as they see fit, guided by their own normative choices, so long as they do not directly bother others. Any lawful action that makes money is implicitly legitimate, however wasteful, tasteless, or morally reprehensible. It is not the job of others to criticize.

This market-encouraged unwillingness to judge the behaviors of other market participants is easily transformed into a reluctance to judge one's own actions. If a specific behavior by another person is not subject to moral censure, then why is it wrong to engage in the same behavior personally? If other people will not judge us harshly for a course of conduct, given the market's neutrality, then why should we be so hard on ourselves? Why not indulge? By similar reasoning (as noted in the last chapter), the market can serve to cleanse goods and services of any troubling aspects about them. How a good is made and at what cost is a matter for the producer to decide, and the producer's behavior is not for other market participants to evaluate. Similarly, waste-haulers are independent market participants, and what they do with waste once it is hauled away is their business alone. Waste-generators and waste-haulers are connected only through a transaction in the market and the market is an amoral meeting place. No questions are asked on either side and no lines of responsibility are forged. It is an industrial culture, Wendell Berry has said, of the one-night stand:

"I had a good time," says the industrial lover, "but don't ask me my last name." Just so the industrial eater says to the svelte industrial hog, "We'll be together at breakfast, I don't want to see you before then, and I won't care to remember you afterwards."[2]

In these ways the market and market-guided thinking tend to undercut cultural cohesiveness and shared moral judgments. They weaken senses of community. And they elevate to primacy the ideal of negative individual liberty. Just as much, by diverting our scrutiny they constrain and confuse our efforts to identify the root causes of land degradation. In a sense, market participants as such are innocent actors, given our unwillingness to judge them. If they are simply pursuing their personal preferences or exercising their negative liberties then they cannot be blamed. The origins of any problems thus must lie elsewhere. If private actors abuse land that they own, then the problem must lie with the lack of good laws or (even better) lack of economic incentive programs that are potent enough to lure landowners into acting better; the problem can't lie with the owners themselves who have acted in ways we can and should condemn. If the commons suffers a tragedy because individual users take too much from it, then the problem is due to the lack of a better institutional oversight arrangement. The unmanaged commons causes the tragedy, not the short-term selfish individualism that propels the individual abusers: hence, the tragedy of the commons, not the tragedy of selfish individualism.

Incentives to Use and Misuse

These observations provide a good background for taking up the particular consequence of the market that one might place first on the list: the ways that the market encourages participants to degrade nature in order to make money and otherwise stimulates overuse.

As one of its much-cited features the market encourages producers to cut their costs in pretty much any way they can. A production process that costs less than alternative means is preferable for that sole reason, even if the low-cost alternative harms outsiders or degrades assets that the producer owns. Indeed, the quest to cut costs can affirmatively encourage a search for ways to externalize harms when doing so facilitates production. With future costs and benefits discounted it can make good economic sense to degrade or consume what one owns, taking the resulting profits and investing them in something with a higher rate of return. Both the lure of gain and competitive market pressures can encourage the full exploitation of nature, focused entirely on commodity production. Thus, a farmer can rip out old fence rows and drain any wet spots to turn every square meter of land over to crop production. A land developer can similarly ignore wildlife habitat or natural hydrologic systems in the quest to wring as much money from land sales as possible. The guiding vision is not the long-term health of the land parcel and the surrounding landscape: it is the maximization in present money of the asset value and revenue flow. In the quest to maximize profits entrepreneurs are commonly encouraged to take chances and risk failure. For many, caution is not a virtue. The market honors a different kind of character: the gambler, the first adopter, the one who acts early and ignores worrisome signals of danger, the one who treats nature and people as objects of manipulation. It is not the personality type who is likely to screen options carefully to ensure that they sustain the communal well-being.

Even as it encourages such misuse of nature, market capitalism and the private property rights that facilitate it are by no means useless in stimulating better land use, as defenders are quick to point out. They can encourage people to keep lands productive. Their good effects are often illustrated using the classic tale of the tragedy of the open-access commons, the tragedy of overusing nature that can arise when competing users of a place are unconstrained in their actions and each is motivated by the lure of gain to expand her use of the natural resource. As each actor expands her use of the natural resources (the grazing common, in the story as famously told by biologist Garrett Hardin), the resource gets overused and declines in quality and productivity. When instead the commons is divided into separately owned shares, the individual grazer-owners have greater incentive not to overuse their respective portions. If they do abuse, they suffer the ill effects themselves. With private shares they are more likely to keep the natural resource healthy.

So far as it goes this argument has merit, and privatization of nature in many parts of the world today might well be (depending on details) a way to reduce land abuse. The limitation with this process is that it hardly guarantees good land use and in some ways even pushes in the opposite direction. To fragment a common resource into individual, bounded shares is to increase the problems (the market imperfections) caused by externalities. Each owner can now benefit by using nature in ways that shift costs to others. Also, secure exclusive rights to use a part of nature can encourage an owner to bring in expensive equipment and technology to exploit nature more quickly than would be possible using mechanisms suited to a weak-tenure regime. The small-scale logging of an unmanaged forest can, after privatization, turn into industrial-scale clear-cutting. Private property, as noted, is not potent enough to sustain the land's health, for reasons long known. How well it does protect the land depends greatly on the particular legal elements of ownership, on how the owner can use the property. These legal elements in turn depend upon the virtues of lawmakers and the effectiveness of law enforcement.

As often pointed out, markets when well functioning encourage producers to become more efficient and reward participants who do so. Efficiency is a much-cited market virtue. The claim, to be sure, has truth to it. But the benefit is sometimes not so clear and even disappears when the overall accounting of benefits and costs takes into account, not just the lower costs paid by the winning producer (the definition of efficiency), but also the costs incurred by losing competitors and outsiders. When a factory cuts costs by moving operations from one city to another, the original home city can suffer economic setbacks due to job losses, declining property values, and business failures. These costs are external to the factory operation, and typically ignored when efficiency is touted. When new box stores enter a market and draw away business from competitors, they might well offer customers better value at lower cost. But what about the competing businesses that have now gone under and the disappearing investments in them? What about the city that now sports a growing number of empty or underused boxes with likely declines also in the values of adjacent lands? What of the customers who must drive by the empty stores to get to the new winning stores? What about, similarly, the simple wastefulness of having multiple gas stations or drug stores on a single intersection—a wastefulness easily avoided in countries with more state-owned operations and better planning?

These inefficiencies of competition ought to be better known than they are. In effect, a city that needs gas stations is itself a kind of open-access commons where competition breeds waste. In the case of an unregulated fishery or oilfield, the rush of competitors to capture the resource before others arrive leads to too much money invested in fishing boats and gear and too many oil wells drilled. The solution to such commons problems is coordinated action at the level of the commons: planning that allocates fishing rights so as to reduce the number of fishing boats; planning at the oilfield level that cuts the number of wells to the minimum needed to extract the oil sensibly; planning at the city level to cut the number of gas stations or drug stores (much as school districts build just the right number of schools). In a community with a single garbage collection service, one truck can go down each street each week. When garbage collection is private, multiple competing trucks bring their noise and gas consumption to each city block. Planning, to be sure, can be done badly and corrupted. But done well, done by people committed to the common good, it is very often more efficient and environmentally friendly.

There is a further concern about market capitalism that has come to loom particularly large in recent decades. This is the long-known need for a capitalist economy to continue growing and expanding if it is to avoid undercutting itself. The whole point of capital accumulation by owners is to generate profits and to plow them back into more capital assets to generate even more profits. Hopes for greater profit are often frustrated, of course, and the process is often disrupted. But the system as a whole works this way and seems to need to work this way. By one estimate, market systems based on privately owned capital need to grow at least 2 percent per year to avoid sliding down. A downward slide can take place when and as market competitors reduce their labor costs—by, typically, shifting to lower-skilled jobs and replacing labor with equipment. As they do so workers collectively have less money to spend. This means lower overall demand for goods and services. For a time the loss of demand can be covered by continued business purchases of labor-saving equipment. But that process comes to an end if demand continues to sag. Unless government steps in to bolster overall demand, unemployment and poverty rise further, leading to yet further declines in demand.

Economic growth, it must be emphasized, is not inherently bad from an environmental perspective although it can be and often is. When the environmental call first went out to end growth—to shift to a no-growth (or, as some now urge, a de-growth) economy—many who heard it reacted harshly without taking time to listen. Scholars who pressed the limits-to-growth claim (Donella Meadows and Herman Daly, among others) distinguished clearly in their writings between economic *growth* and *development*. Growth as they defined it meant continued increases in the consumption of resources taken directly from nature and in the demands made on the planet to assimilate human wastes. That is what needed to end and be reversed, they said. What they did not oppose, and in fact strongly supported, were qualitative changes in economic activities that brought enhancements and higher levels of goods and services but did so while decreasing the consumption of raw physical inputs (by using inputs more efficiently, greater reuse and recycling, shifts to renewable energy, public transport to replace private, and much more). This kind of expansion they termed economic development, and it was good. Development should continue; growth needed to end.

Decades later the no-growth proposal continues to inflame opponents who either do not understand it or find it helpful to misunderstand. In the meantime, the proposed definitional distinction between growth and development has not caught on. Calls to improve the economy and maintain employment continue to blend together measures that would expand ecological degradation (growth) with measures that would not (development). Without greater limits imposed on it from the outside, the market will continue expanding its consumption of nature as an apparently inexorable part of basic operations. More wetlands will be drained and filled, more grasslands plowed up, more and newer chemicals dumped into waterways, more aquifers drained down, more species pushed aside, more forests converted to tree farms, and more greenhouse gases emitted. To top it off, the miscast modes of seeing and valuing that are promoted by the market will make it harder for people to grasp what is going on: to evaluate it, to identify the root causes, and to take collective action for change.

Sapping Communal Power

For reasons already mentioned, the capitalist market based on private property tends to discourage public evaluations of choices made by producers and consumers. It tends to accentuate the cult of objectivity and to push questions of moral value into the private realm except in the case of values that have long been embraced at the public level (chiefly: the moral value attached to human life, the ideals of liberty and equality, and the benefits of public safety and economic growth). These traits of market capitalism tend to diminish senses of community. They weaken support for collective action to promote new values. Other aspects of market capitalism also push in the same directions, further weakening communal bonds and sapping the strength of democracy.

The modern mind, shaped by market thought, has trouble coming up with a vision of good land use and thus trouble evaluating land uses that are taking place. Similarly, the market-shaped mind contains no real sense of community welfare or the common good and is thus slow to talk about it. Much questionable land use is undertaken by private producers on private land. The market tends to shield their production methods from scrutiny, just as cultural ideas about private property leave owners free to use their lands as they see fit so long as they don't harm neighbors. Private land use is the owner's business, not the public's. Meanwhile, as it fosters lower wages, envy, and greed, the market creates demands for ever-higher levels of growth, particularly short-term. All of these forces resist long-term, sensible democratic planning.

Democratic action also slides down the more people think of themselves as separate beings who can and should go it alone in the market. It slides as normative values, for the reasons mentioned, get pushed into the private realm of consumption. Actions by market participants as individuals are subject to what has been termed the tyranny of small decisions—the bad consequences that arise as individuals make choices that are sensible for them as individuals but that drag down the collective whole when many people engage in them. For one farmer, starting irrigation can make sense; when too many do it, the aquifer is drained. A small amount of nitrogen running into waters does little harm; too much creates vast dead zones. An extra few apartments in a neighbor might have little effect; when dozens of developers all build, the civic infrastructure can be overwhelmed.

In the worldview of the market, the way a person gets something is by gaining money and then going out to purchase the thing wanted, not by working through government channels. From that perspective a person who doesn't like sprawling development can get together with like-minded people to purchase and hold conservation easements. Someone who cares about animals can give money to animal shelters to care for them. Those who value clean air can move to a place that has it. In the market view, every desire is simply a consumer preference that the market can handle. Those who want to change the world are thus instructed to convince other consumers to share their preferences and to use their purchasing power to bring it about. So strong is this market message that when people ask—what can we do?—they typically mean what can they do as consumers; what can they do as individuals acting alone and with friends?

The core problem here is that, in the costly worldview of the market, people are chiefly participants bound together by the amoral market and free to act on their preferences as individuals. They are not, in contrast, citizens who are bound together by their membership in local, state, and national body politics and capable of acting collectively through governmental means. The roles of citizen and consumer are often starkly different and the preferences of individuals can vary considerably depending upon the hat that one wears and thus the options available. As citizens we are invited to think about the common good and about laws and policies to promote it; the available options now include joint action. As consumers, in contrast, we are invited to think selfishly. As citizens we are parts of something larger and work together. As consumers we are autonomous actors who go it alone. The more the market dominates, the more we are pressed into consumer roles, the weaker democracy becomes.

These many signals and images embedded in the market become all the stronger when the market is supported (as it is) by mythology about its vast powers. Market forces, to be sure, have given rise to exceedingly clever inventions. The market has shown a remarkable ability to find solutions to technological problems when there is money to be made or when regulators demand it. Its impressive record leads people to place great confidence in its ability to solve all manner of problems as they arrive, without need for fundamental changes in other realms. Given what the market has done in the past, cannot we trust it to find solutions to new problems caused by ecological degradation? Cannot we charge ahead, emitting more greenhouse gases, confident that market-driven scientists will find ways to remove them from the atmosphere? Cannot we degrade soils confident that we can produce food in other ways, or destroy wild fish populations knowing that fish-farm operations will arise to take up the slack?

Faith in the market's cleverness provides a reason for resisting democratically driven public laws and policies to address ecological ills. It lulls people into inaction. At the same time, it convinces people that individual liberty, respected by the market, is not inconsistent with good environmental outcomes. The market's commodification of nature is similarly not a problem, nor are extensive rights for private property owners. All are fine because they are embedded in a market system that, on its own, will find solutions to problems. It is a popular stance, one that inhibits collective action through democratic means while at the same time clouding the search for the root origins of land abuse.

Finally, there is the troubling reality that concentrated wealth, arising in the market, has gained so much power over politics and law-making processes. These days, candidates for office are beholden to wealthy donors, who increasingly screen candidates based on their pro-business stances. Legislators in the US capital are greatly outnumbered by business-supported lobbyists seeking to influence them. Regulatory agencies are hounded by them and their lawyers. Businesses tend strongly to resist limits on their operations and push to cut law enforcement budgets, except insofar as they can turn legal limits into opportunities for monopoly profits. Businesses also tend strongly, when possible, to take over commons assets and turn them into private property (for instance, gaining intellectual property rights in the products of publicly funded research). Increasingly, governments are cast in the role of advocates for local businesses, helping them compete in global markets and offering incentives to keep local jobs in place. Inevitably such advocacy—taking the side of business and using government to help it—creates tensions with the essential role of government as a counterbalance to concentrated wealth and as the democratic means by which the ill efforts of markets and competition are kept in check. Then there is the willingness and ability of many market actors to use concentrated wealth to distort public understanding and discussion of key issues. Oil-company funding of climate-change denialists illustrates the illness. So do claims by agri-business that its manipulation of genes and high-chemical land-use practices pose no health threats.

These various messages of the market, the mythology that surrounds it, the deliberate business efforts to take over government: alone and in combination they greatly weaken citizen efforts to address collective problems. When it comes to democracy, when it comes to citizens using government to improve their world, the market is a massive and still-growing opponent.

As for the effects of this distorting power, they appear in a recent study by political scientists Martin Gilens of Princeton and Benjamin I. Page of Northwestern. Their inquiry, drawing upon researches of their own and by countless others, considered how influential ordinary citizens were in setting US policies at the federal government level. Their conclusions are stark. Average citizens get their way only when their preferences are also supported by economic elites and by organized groups representing business interests. When average Americans support a public policy that is opposed by elites and business groups, their preferences "appear to have only a miniscule, near-zero, statistically non-significant impact upon public policy." Mass-based citizen groups, environmental groups included, also seem to have little effect on public policy when they too promote stances opposed by economic elites and business interests. Compared with them, "business groups are far more numerous and active; they spend much more money; and they tend to get their way." The bottom line, for Gilens and Page, is alarming:

When a majority of citizens disagrees with economic elites and/or with organized interests, they generally lose. Moreover, because of the strong status quo bias built into the U.S. political system, even when fairly large majorities of Americans favor policy change, they generally do not get it.[3]

All of this, needless to say, bodes ill for democratic governance and collective action. It bodes ill for people who take off their consumer hats and put on their citizen hats. The failings here can be attributed to bad politics, and they are linked to that. But the system as a whole arises out of and stands atop a cultural order that in important ways supports and sustains it. It is a cultural order that exalts the individual and liberty, one that has more faith in the market and its fairness than in politics. It is a culture that sees the world in fragmented terms, that expects and tolerates aggressive competition, that often favors competition over cooperation, and that is content with clear winners and losers. Nature as such merely feeds into the market, supplying raw materials, and when materials run short the market is expected to find substitutes. Communities as such, both social and natural, are of little importance and can rise and fall as they will.

Democracy has declined in recent decades, organized citizen power has declined, in large part due to cultural values and prevailing ways of seeing the world and situating individuals within it. It is hardly coincidental that these values and ways of seeing the world are embedded in the market and in the larger market-centered view of the world. The same values and ways of seeing are among the root causes of land abuse, which the market on its own so readily facilitates. Efforts to deal with one of these problems, if successful at the root level, are likely to bring gains to the other one as well.

Tinkering versus Cultural Change

The environmental reform movement today includes a growing number of voices calling for changes to the economic systems, calls for a new economy. Milder versions of the call look for such changes as campaign finance reform and for regulations that better control the eco-logical ills of business activities. Stronger versions take aim at corpo-rations themselves and want to see changes in them, whether simply greater transparency in what they do or some direct voice in corpo-rate governance for community advocates and perhaps labor. Nearly all want to see new ways of measuring the health of our economy and the nation, using standards that pay attention to much more than just volumes of market transactions. If our economic indicators, so often broadcast, took into account environmental harms, then perhaps we would become more aware of them. For many, a direct target of attack is the US Supreme Court and its growing list of rulings that elevate corporate powers and constitutional rights. Citizen democracy is being caught in a powerful pincher movement—with concentrated wealth taking over government offices while the Supreme Court, on the other side, saps government of power to regulate business. One need not look ahead very many years to see a time when government is largely con-trolled by business, particularly at the national and state levels. The two forms of power will become one, and work in tandem.

Reformers have also long talked about practical ways of reducing market incentives to engage in misuses of nature. Here we have calls for shifts in the overall approach to taxation, taxing pollution and the consumption of raw resources rather than, as now, taxing income and labor. Calls continue for new regulations that internalize the externali-ties of business activities; for instance, making land users (farmers, de-velopers) accountable for the water pollution that runs off their land, much as factories have been for several decades.

Such changes offer the prospect of mitigating current ills. But do they offer hope for more substantial improvement? Or to rephrase the question, are they aimed at the true heart of the problems posed by the market?

In many ways, the market reflects who we are, as defenders some-times point out. It operates to provide us with what we seek. If our pre-ferences are ones that, when fulfilled, degrade the natural world, then we ought to blame ourselves. If we would only want different things, only seek different goods and services, then the market would lead to better outcomes. Don't blame the market, defenders say: It is merely a tool, and the best one we have to promote overall economic welfare.

This defensive reasoning contains merit, both in pointing to individual preferences and in noting how the market can in fact help achieve a wide variety of goals, including goals consistent with land health. But the reasoning also has severe flaws. Chief among them is the implicit claim that the market can be thought about separately from people, as some sort of neutral mechanism into which we plug ourselves (as producers and consumers) and nature (the stockpile of natural resources and waste-assimilation capacities). The reasoning suggests that preferences are formed apart from the market (they are exogenous, as economists put it), which is to say they are not stimulated, shaped, and constrained by the market. And the reasoning avoids comment on how this affects people in their capacities as citizens, just as it ignores how so many possible actors—communities as such, landscapes as a whole, future generations, other life forms—are not participants in the market. Only individuals with money to spend or services to supply are allowed to play the market game. It is a game that strongly favors some outcomes over others, that rewards some personality traits over others, that pushes people to see and think in certain ways rather than in other ways.

The central problem with the market—or more aptly, the large constellation of problems at the center—has to do with the worldview that it embodies and implements, strengthens, passes on to new generations, and wraps in an aura of inevitability. The market's chief effects, that is, are not directly on the land itself, not on nature and its functioning, but instead on the minds and hearts of people. Yes, the market is us. But having created the market it takes on a life of its own. It is not merely the passive product of people living today, the result of who they are and what they want. It has much greater agency than this. It is a worldview that jealously repels alternatives. Much as a new technology, once developed, can assert power by inviting us to use it in new ways—ways that shape how we think and feel—so too the market, once alive and at work everywhere, greatly affects what we see, how we understand ourselves, what we value and want, and much more.

Because of this, the market cannot be altered fundamentally simply by tinkering with details in the ways that most new-economy advocates have in mind. One reason for this is that the market is powerful enough to resist such reforms. The reforms will not seem sensible to people within the worldview that the market presents, the worldview that is now so dominant. The problems that reformers point to—ecological degradation among them—will be hard to see, they are hard to see, for the same reason: the hegemonic dominance of the market perspective. The reforms themselves will seem too costly, again when considered from the perspective of the market view, the one that exalts individual liberties, that pushes preferences into the private realm, that sees the world from a short-term perspective, and that is willing to gamble. Etcetera.

We are failing at our oldest task due in very large part to flaws in our culture. The flaws affect us directly. They affect us perhaps even more because they are so entrenched in the market and have, in that form, taken on a life of their own. At the risk of repetition it can help to re-iterate and assemble the pillars of this powerful ideology so that it can be seen as a whole:

· Humans differ in kind from all other forms of life and rightly dominant on the planet.
· Nature is physical stuff, objects here for us to use as we like.
· Nature comes in the form of fragmented pieces and parts; interconnections and functional processes are of little importance, as are organic wholes.
· People are basically autonomous individuals, independent actors, and any com-munity roles (social or natural) are optional.
· People are market consumers and producers (service providers), not chiefly citi-zens; even when voting they are chiefly consumers.
· Future generations and other life forms, like landscapes and ecosystems, play no direct roles; they show up in thought and policy only insofar as individuals choose to include them.
· The public realm is guided by instrumental reasoning, using facts and logic; val-ues, norms, aesthetic and religious preferences are pushed into the private realm for consideration by individuals whose choices should largely go unquestioned by others.
· Market actors are free to do anything they can to make money, setting ethical concerns aside, subject only to legal limits that are enforced with enough sever-ity to make it economically sensible to abide by them.
· The way to get something is to buy it or manipulate market elements to obtain it, not by working as citizens.
· The proper time horizon is short term and rewards rightly go to those who take risks and charge ahead, not those who hesitate and worry.
· The market can be relied upon to harness human cleverness and solve problems when it is really necessary to do so.
· Acts that make money, acts undertaken by private landowners on their lands, are presumptively acceptable if not by definition efficient and thus desirable. To the extent they seem to entail ecological harm the claim of harm is often mis-leading if not erroneous because the market has ways of turning apparent vice

into virtue. In any event, income-generating activities contribute to the welfare and help fund environmental improvement efforts.

The creative powers of the market we hear about often and the evidence of it is pressed by marketers twenty-four hours a day every day of the year. The ill effects of the market, on people, land, and the public realm, are hidden from view, sometimes because of their remoteness, sometimes because the evidence of harm is invisible to ordinary senses, and quite often because unclear and ill-based normative thinking undercuts the ability to evaluate sensibly. As patent ills nonetheless mount, particularly in terms of unemployment, poverty, and gross inequality, market thinking and market defenders often stand ready to divert the resulting anger and frustration onto government, further weakening prospects for citizen-led reform. The market's sole competitor for generations, the sole power able to contain the market, has been government. To weaken support for it, to sap it of its powers, is to leave the playing field to the market alone.

Reform of the economy is not rightly thought about as a separate undertaking. It is better understood as part and parcel of a larger effort to bring changes in dominant culture, particularly cultural elements that enliven the market and enable it to perform as it does, that enable it to retain such hegemonic control over the modern worldview. For this reason, targeted reforms seem unlikely to get anywhere, not anywhere substantial, unless this worldview is challenged directly. For that to happen, for people to start seeing and valuing the world in different ways, they need to start hearing, time and again, different ways of thinking and valuing. They need to be presented, again and again, with new frames that clash with assumptions embedded in the market view. It is not enough, however, simply to push forth specific ideas in isolation: the claims, for instance, that more parts of nature should be covered by public trust duties, private lands included, or the claim that communities should gain voices in corporate boardrooms. The ideas are good ones, but to put them forth in isolation is much like sending a lone soldier against the entrenched machine guns of any enemy: the valor is there, the cause is good, but the machine guns will win.

Fundamental cultural change requires an overall strategy, well considered. It needs a strategy that identifies all of the pieces that need reform and that assembles them in a good order. That way, progress on one element helps promote the next and no element is put forth ahead of its logical time.

Beyond Civil Rights

When environmental reform is brought up for discussion, a common response is that the environmental cause should draw more lessons from civil rights causes, particularly, given recent history, from campaigns to promote gay rights and gay marriage. That reform effort has made strides in the face of distinct headwinds. Cannot the same methods used there be borrowed and used by the environmental effort? The question is good, and the answer can be useful.

The gay rights campaign has not had to do battle with the powerful market and has not pressed for any changes to the market or to any important element of the market worldview. To the contrary, gay rights is entirely compatible with the market because it represents, in market terms, just another constellation of individual preferences that the market is ready and willing to satisfy. The market's version of equality is simple: one person's money is just as good as another person's. As for market participants who discriminate against gay people, they are simply losing chances to make money. Other suppliers will be happy to take the business.

Beyond this, the central moral claim behind gay rights is very much in line with the values embedded in the market. The market is all about human exceptionalism, autonomy, and negative liberty. It is about people getting to make their own normative choices and keeping moral values away from the public realm except as needed to keep the peace and protect the market's functioning. Gay rights has nothing to do with nature and how we perceive it and value it. It does not insist on any sort of community-upholding, organic view of society or the world. In Michael Sandel's terms, it is morality based on freedom, not on virtue. To be sure, some gay-rights rhetoric has promoted communitarian-type goals. But the bottom line of the efforts is that gays should be able to live as they like and forge links as they choose. The market is entirely happy with this.

More generally, the gay-rights cause is very much in the tradition of liberal individualism. It builds on values (liberty, equality, property) that are already deeply entrenched, asking only that they be applied in a slightly different context. It calls for no sacrifice by anyone. It imposes economic costs on no one. It does not threaten private property rights. It does not interfere with efforts to grab monopoly profits. It does not weaken the tightening grip of economic elites on government, including the Supreme Court.

In sum, the gay-rights cause bears virtually no resemblance to the effort to promote good land use. The obstacles it has faced, though sizeable enough, were and are modest in comparison to those faced in the environmental arena. The cause was an easier one also in that single actors—state governments, one by one—could largely give to gay advocates what they sought: legal recognition. The cause, of course, has also pressed for social acceptance, a related but different and higher hurdle. And that part of the effort, changing public understandings, does provide a lesson that environmental advocates might draw upon. But that lesson needs to be kept in its relatively small place. The environmental reform effort needs to present, not a single, easily phrased message (as the civil rights movements, in every form, could do), but a new worldview, one with many components. An attempt to simplify the message—down to, for instance, "be nice to nature" or "love Mother Earth" or "care for future generations"—will lead to the same kind of frustrations and failures that have marked the cause for the past quarter century.

Major cultural change is needed, of a type far beyond that of any civil rights cause, change of a type that has, by way of earlier examples, perhaps only two precedents: the shift 10,000 years or so ago from the hunter-gatherer world to settled occupation and agriculture, and the shift beginning in the seventeenth and eighteenth centuries to an industrial, market-based worldview laced with liberal individualism. A similar sea change in mentality and sensibility now seems necessary.

EIGHT

The Path Ahead

Our misuses of nature are not due chiefly to any lack of information, whether about nature or about how we're altering it. Nor are they due especially to any lack of green technology, although better technology, like better facts, might well help. A rising world population, to be sure, does add stresses to the planetary system. But population is by no means the chief problem (rising per capita consumption is more acute) and populations in many regions have stabilized. The major culprit, the central cause of our misguided acts, is modern culture broadly understood. As we've seen, culture includes the ways we perceive nature and make sense of it, the ways we value what we see, and the ways we understand our place in the natural order. It has to do with our moral orders, our time frames of understanding, and our confidence in our cleverness. We shouldn't be surprised that our cultural trajectory has led us to where we are, that it hasn't developed in ways that pay proper attention to today's environmental ills. Our evolutionary trajectory as living beings did not adapt to them because there was no need. Our developing moral ideals have also not embraced these problems, not as we need them to, mostly because morality develops slowly and our environmental ills are comparatively recent. But moral order evolves. Culture has and does change.

A more land-respecting culture can in fact emerge, including sounds ways of seeing and valuing nature. Further, the path leading in that direction, if people could only look down it, should offer much that is appealing to us. To move ahead, though, a major reform movement is needed, a movement that seeks nothing less than revisions in our culture's trajectory. This will involve something far different and grander than the well-known civil rights reform efforts of recent generations. For starters, we need to recognize the nature and scope of the vital work that lies ahead. We have not yet done so, and reform efforts have stumbled because of it.

In a perceptive recent study, *The Age of Fracture*, Princeton historian Daniel T. Rodgers traces the many ways American society after World War II became more fragmented and individualistic. Over the era society gave decreasing attention to social ties, context, and common interests with a greater commitment instead to self-selected, fluid identities and to a worldview dominated by the market and competition. "Strong metaphors of society," Rodgers reports, "were supplanted by weaker ones." "Conceptions of human nature that . . . had been thick with context, social circumstance, institutions, and history gave way to conceptions of human nature that stressed choice, agency, performance, and desire." What changed most of all, Rodgers contends, "were the ideas and metaphors capable of holding in focus the aggregate aspects of human life as opposed to its smaller, fluid, individual ones."[1]

This change in tone and cultural focus showed up clearly in the shifting presidential rhetoric between the Carter years and those of the mid- to late Reagan years. While he was president, Carter "talked easily of humility, mercy, justice, spirit, trust, wisdom, community, community and 'common purpose.'" He carried forward in doing so language used often by his predecessors. There was John Kennedy's famous admonition to "ask not what your country can do for you. . . ." There was President Nixon's contention that no man was truly whole "until he has been part of a cause larger than himself. "To go forward at all," Nixon had said in his First Inaugural, "is to go forward together." Once settled in office, President Reagan shifted to more self-centered rhetoric. He encouraged individuals to pursue self-actualization, to imagine how they might get ahead separately. "Reagan's word-pictures of the people," Rodgers records, "almost never showed them working together, their energy and talent joined in common action." Reagan pioneered the use of the personal story of the individual rising up against adversity, often pressing against society and government. "In Reagan's very celebrations of the people, the plural noun tended to slip away, to skitter toward the singular."[2] A well-guided effort at environmental reform will need to propel our shared thinking, our shared culture, in a different direction.

Our Cultural Deficiencies

Contemporary public culture—putting aesthetics and tribal loyalties to one side—intermingles three major cultural components. All have played roles in earlier chapters, but it is useful to invite them back on stage, to see how they work in tandem and to reveal how their limitations account for and track the flaws in modern culture.

Perhaps the major building block of current culture is the constellation of ethical values and understandings long embedded in Christianity and pushed forward in Western society by the church and Christian writers. As Oxford historian Larry Siedentop charts in his insightful overview, *Inventing the Individual*, it was Christianity—with borrowings from ancient Greek culture—that raised high the individual human as a morally worthy being, created in the image of God. As it did so Christianity challenged earlier moral orders that embedded people into families, clans, and tribes and that understood them chiefly as parts of organic orders, which were themselves the chief repositories of value. Over the centuries, Christianity slowly called into question these organic orders and the hierarchies and wide status-differences that they endorsed. Implicit in the new religion was an ideal of equality, reflecting the basic moral importance of all individuals. Individuals as such counted, the church said, which meant, in time, that individuals as such—contra earlier thought—were properly understood as moral agents. An individual could walk away from her family and take up the cross; an individual as such could leave home and join a monastic order. Judgment and salvation came to center on the unique personal soul. Compassion and respect were the new guiding ideals. The family as such counted for much less; moral identity was no longer intertwined with civic membership or engagement.

Christian social ethics brought vast gains, to be sure. But it needs noting that Christianity, for all the moral value that it put forth, presented a quite constricted moral vision overall, one that dealt almost exclusively with one-on-one dealings among people. Christian morality did not offer a plan for a just, durable society, or anything close to it. It did not talk about the welfare of communities as such. It said nothing directly about the common good, except by highlighting how society should respect individuals and honor their equal moral value. Nature was not even in the picture and it certainly no longer contained hidden spirits. By implication only humans possessed moral value; they were best understood as autonomous individuals; and future generations counted for little. Christianity certainly encouraged individuals to share their wealth and aid the poor, but it also encouraged individuals to look after themselves and their futures, opening the way for cultural shifts that fostered self-concern, hard work, and affluence.

This Christian social order, in both its religious and (increasingly) secular garb, played a key role in the origins of the second major component of today's public culture, the component having to do with political rights and, by extension, individual rights generally. The notion of human rights dates back earlier than commonly thought. Siedentop embeds it in the High Middle Ages with the writings of church scholars on natural law and, later, natural rights. As already considered, rights-based thinking in secular form stood tall in the eighteenth and nineteenth centuries, mostly in political realms. Particularly as expressed in the US Bill of Rights and related iterations, individual rights had to do with how the individual fit into the body politic. Rights were largely procedural in the sense that they played roles in new systems of popular governance. The aim was to create a political engine that would avoid oppression and, when possible, work for the common good (the "general Welfare," as the US Constitution put it). This rights-thought did not propose any particular vision of the common good. No more than Christian social ethics did it include an image or plan of a healthy community as such. It too had nothing to say about nature and human dealings with it. It too recognized moral value only in humans now living and presented them as autonomous beings. Rights-rhetoric, in short, was offered as one piece of a new moral and cultural order, a narrowly drawn solution to the political crisis of the day.

For generations Christian social ethics and the language of political rights worked in tandem. They were adequate to keep society moving along, particularly when strengthened by inherited ideas of individual virtue (honesty, integrity) that sank older, secular roots. Indeed, the two cultural components together supplied an impressive array of intellectual and moral tools that social and political reformers could put to good use, particularly civil rights campaigners. And they did so, even as public arenas became more secular in form and as the Christian origins of social ethics were veiled or forgotten.

What we now can see, what we need to see, is that these two cultural inheritances, the centerpieces of modern public morality, rest on problematic assumptions and have vast gaps. They are distinctly human-centered, with little room to recognize moral value diffused in other life forms and communities. The Jewish scriptures—the Christian Old Testament—included passages that honored God's creation, but core Christian teachings made little room for them. Both moral orders, as noted, focused on the present alone, and both exalted the individual as independent moral agent. Had political rights been kept in their proper place—as safeguards to ensure good government—they might not have gained the great influence that within a century they did. But they expanded their reach as economic elites, in their efforts to ward off government controls, began to contend that the protection of these rights was a chief end of good government, not merely a procedural component of it. The common good, that is, became increasingly defined in terms of respect for individual rights, particularly liberty and private property.

Social ethics and political rights, then, worked well to drive various civil rights campaigns and related initiatives to help individuals. But as a repository of intellectual and moral reform tools, the combination had almost nothing to offer to help interpret and remedy looming environmental ills. Worse, with the dominance of social ethics and political rights in the public realm—and by the late twentieth century they had certainly become dominant—little room remained for other moral ideals to squeeze in. Moral value, again, seemed limited to individual living humans. On the political side, the rising emphasis on individual liberty cast grave doubt on the legitimacy of health and safety regulation. Indeed, to respect liberty in its various forms (particularly private property) was to call into question most steps that a government might take to promote the common good, including healthy landscapes. Similarly, a stress on liberty greatly affected ways of dealing with human ignorance and nature's incredible complexity. A virtue-based approach could build on humility and embrace a cautionary attitude, minimizing changes to nature when consequences were unknown. A liberty-based approach, in contrast, insisted on a much different stance. With liberty held high, individuals as such—the business owner, the property owner, the wealthy consumer—were entitled to charge ahead, acting as they saw fit, unless critics could offer clear proof of looming harm. Rights largely trumped virtue. Ignorance offered a green light, not a yellow or red one.

Both social ethics and political rights, of course, have had their advocates. And they retain respect, both because important truths and values lie within them and because their track records are impressive. But to admit this is not to refute the high costs that have come by giving them such dominant roles in the collective moral order. They do not, as reviewed at length, supply adequate tools to distinguish the legitimate use of nature from the abuse of it. When it comes to talking about our natural homes and good living, they constrain the talk, question its public legitimacy, and push it into the realm of individual choice—not fully or successfully, to be sure, but with great effect. At the same time, their dominance and aura make it hard for us to identify the root cultural causes of land abuse. And to the extent observers point toward cultural flaws, they are readily pushed aside. A common defensive move is the claim that the values critics promote (moral value in other species, for instance) are simply matters of personal choice, ones for individuals to embrace in private life. When that doesn't work, the bigger defensive guns are rolled out: Critics are condemned, in so many words, as moral traitors—as misanthropes, socialists, and agents for overweening government.

Social ethics and political rights, in short, do not merely dominate the public moral arena. They go further to resist inroads by new ways of moral thought. They are, one might say, jealous of their hegemony.

Standing along with these two cultural components is a third one, every bit as important and overbearing even as it modestly veils its moral influence: the capitalist market. In key ways it has come forward over the generations to fill-in major gaps in social ethics and political rights, particularly by structuring the ways we perceive and value nature. In doing so, significantly, it has also pushed hard against both social ethics and political rights so that its realm and power continue to wax. In tandem with them, the capitalist market has added to the current difficulties faced by green interests and other moral reformers.

The moral grounding of the capitalist market overlaps considerably with the pillars of social ethics and political rights just described. It too, as we have seen, honors the individual as such, the consumer and worker. It exalts not the morally equality of Christianity or the political equality of the Bill of Rights but rather the market's willingness to accept all people, regardless of traits, based on the money and skills they have on offer. Individual agency is respected—putting to one side the grave inequalities of economic power—and liberty is barely restrained. Where the capitalist market goes further is in supplying ways to think about nature, as a warehouse of fragmented commodities. The market also gives content to the common good and "general Welfare" with its emphasis on overall economic activity, measured day by day. Market activity, in turn, is linked in theory to the satisfaction of individual preferences, so they too play a role, providing home for all manner of individual moral quirks. The market thus has its own power-tool to ward off alternative moral values and visions: it pushes them all into the category of individual preferences, where they cause little disruption. In the view of market defenders, the market is not just consistent with a democratic ethos but the apotheosis of it—putting to one side, of course, inequality in economic resources. And if government would simply get out of the way, then the market could similarly exalt individual liberty.

When economic productivity stands as the central pillar of the common good, it becomes yet harder for environmental critics to put forth alternative moral visions for collective rather than individual embrace. It becomes harder, indeed, even to talk about collective values and goals in language other than that of personal preference. In this way, the market defends and keeps secure the primacy of individual moral worth, equality, and political liberty. Yet, even as the market seems to respect these core moral values it quietly but consistently works to cut into them. The market honors greed and purports to turn that vice into virtue; as it weakens longstanding moral criticisms of greed and gambling, the realm of social ethics contracts. As it exalts free choice the market implicitly calls into question longstanding moral condemnations of such choices, particularly by the wealthy. If liberty and equality mean letting people make their own choices, then who is to second-guess the choices people might make? Indeed, when the public good equals high economic activity, then those who produce and consume lavishly are honored market participants, if not immune from moral criticism then at least protected by a sturdy defensive shield.

Added to the mix here is the message that the market cleanses. Market participants bargain at arms-length and are responsible for their personal choices, and only their choices. Consumers thus need worry neither about the costs and means of producing their goods nor about the ill effects of waste disposal. On the other side, producers who simply respond to market forces—who do what other producers are doing—are often similarly absolved of guilt. As for Christian sharing, it too gets pushed into the realm of individual preference; it is not, or no longer, a moral principle to guide action in the public realm. Once recast as an individual preference, of course, Christian sharing largely ceases to provide grounds for public moral criticism. Even with respect to private actions, selfish behavior is protected by the shared norms of individual liberty and individual respect. The wealthy magnate who shares nothing is not of course fully insulated from all criticism: Virtues still carry weight, as do Christian scriptures. But the moral criticism is deflected by the principles of free choice. The same story unfolds in the case of land abuse, similarly shielded from moral criticism by liberty, equality, and private property.

Just as it erodes the realms of social ethics and virtue in these ways, so too the market cuts into the practical reach of individual political rights. Most evidently (as noted in the last chapter) economic elites are slowly taking over governments and manipulating them to their advantage. Constitutionally protected property rights have long been used as means to curtail the public sphere; for instance, public downtown arenas, once the classic arenas of free speech, have been replaced by private shopping malls where the First Amendment does not apply. More recently, claims of religious freedom by the wealthy help insulate private practices (discriminatory hiring, for instance) that infringe the rights of others.

Step by step, so it seems, the market and its embedded views of the world are taking over more space on the public stage. They are squeezing out both social ethics and political rights and, in the process, making it ever harder for advocates of new moral orders to gain traction.

———

In terms of the weaknesses of this overall moral landscape, taking the three components in combination, a quick summary seems useful.

· As repeatedly noted it is an overall moral order focused on individual humans living today, with no real recognition of broader morality in the world except as individuals want to recognize such value in their individual affairs.
· The market alone, with its GDP worship, provides the only real measure of the common good (public safety aside) and the only measure too of how we ought to perceive and value nature.
· Ecological interconnections and interdependencies do not register, any more than do large-scale landscapes (the Mississippi River basin as a whole, for instance) that cannot be bought or sold.
· One can ransack these moral components and still have no good tools to distinguish land use from land abuse, save in the grossly inadequate language of individual preference and economic efficiency.
· Future generations and other life forms are also drawn in, barely, in just this tenuous way.
· Armed with the language of liberty and private property, land abusers can insist that allegations of harmful conduct be supported with vast if not unimpeachable scientific evidence, virtue be damned.
· In this restricted moral order it becomes ever more confusing to make normative sense of nature's dynamism: if nature itself changes, how can green groups claim that human-caused changes disrupt some moral order?

· Similarly it becomes hard to respond to contentions that the market will harness human cleverness and solve problems when and as they arise. To challenge this presumption is to propose and prove a negative, always a formidable task.

Then there are two matters that perhaps rise above the others. One is the current tendency to accept and embrace competition as the proper means of social interaction in economic matters, including most dealings with nature. The market induces and rewards competition and self-seeking behavior, viewing nature instrumentally as raw materials. Political liberties provide no brake on this behavior. Social ethics could protest but they have little effect once confined, as they largely are, to the private realm of individual choice. To honor competition this much is to call into question all other methods of collective decision-making. It is to cabin individuals within their roles as market participants—as consumers rather than citizens—thereby foreclosing critical collective options.

Atop these factors is the matter of nature's limits: the fact that our physical planet, though daily bathed in energy, is defined in size and physical resources; defined in terms of overall water flows and waste-handling capacities; defined in terms of the ways that it operates through ecological processes; defined in terms of its life forms. The market admits of no limits; it consumes and exhausts whenever profit beckons. Private ownership usefully checks the ravish-and-run mentality, but only partially and only when owners by choice act well. As for the other two components of our moral order, they are even quieter when it comes to nature's limits. Social ethics is about sharing what we have today; the religious language of stewardship, of caring for the long term, is a recent add-on and fits in little better than environmental concerns generally. Individual political liberties, far from resisting competition, are largely consistent with a culture and political order that saps nature with alacrity.

This, then, is where we are now. This is the scheme of public values that we need to change, as the central, overarching focus of reform efforts. As explored below, a key reform step is to demote the market from its status as moral co-ruler and embed it in a sound moral and ecological order. The market needs to stay; market-based tools can be of value. It is the market as moral arbiter that needs revision, the market as controller of nature, the market that displaces citizen decision-making and pushes so many moral values into the private realm. As for social ethics and political rights, they do, of course, need to keep important roles. But these moral components too need to share the stage with new understandings and moral visions.

Use and Abuse

Earlier chapters explored at length the varied normative factors relevant in drawing the vital line between the legitimate use of nature and its abuse. The line is one that will need drawing and redrawing in particular landscapes, acknowledging local conditions and needs and eliciting the best normative thought local people can bring to bear. For reasons noted, the relevant factors are best considered and applied at varied spatial scales. It isn't possible to look at a single field or forest and make final judgments about the good use of it. A parking lot, all paved, could rightly qualify as the good use of nature when properly considered at various large spatial scales, if in fact it helps meet important human needs and is situated so that it does not interfere with achieving other values.

Good line-drawing will be done, as noted, on an all-things-considered basis. It is not sensible or useful to validate a land use simply because it helps promote one relevant factor of good land use when it does nothing to promote, and by exclusion tends to undercut, other relevant components. The mistake is far too common: we see, for instance, that a field grows food, note that we require food, and then stamp our approval. It cannot be this easy.

Reformers need to recognize and insist that this line-drawing work unfold as public business, subject to collective decision-making, even as landowners are left free to choose among healthy options. Nature is, after all, our common home. And it is work that involves normative choices. Yes, scientific facts are essential. But science taken alone, for the reasons covered, cannot alone draw this line well or even at all.

Sustainability, as noted, is no more than a feeble step as a line-drawing effort. Let us term it an initial foray, a youthful try, to be followed by more mature ones. The call to promote ecosystem services is a little better, but only a bit. It does arise out of a normative claim that we ought to keep nature productive, and to that extent is good and sound. But the moral content here is quickly clothed in science-sounding language, and then subjected to detailed scientific research and economic number-crunching, leading to number-laden reports intended to show why it makes good economic sense today, in GDP terms and for living individual humans, to keep nature productive. Advocates of ecosystem services know well that good-functioning systems will also help other life forms and benefit future generations. But it sometimes seems that these points are kept quiet as if not really pertinent, or as if they were incidental benefits that we gain when and only when the protection of ecosystem services makes sense without them. In too many ways, ecosystem services, like sustainability, comes across as a variant on the sustained-yield language of the Progressive Era. It is the old idea in slightly better garb: Nature exists merely to meet our needs, we can understand it in instrumental terms (empty of moral value on its own), and we can plow ahead based on facts and reason with little recognition of our ignorance.

The major problem with ecosystem services—even as it represents a real step forward—is that it doesn't challenge cultural values as such. Only indirectly does it hint that we need to see and appreciate nature in different terms. It does not address basic questions about the diffusion of moral value in the world. It puts itself forward as a good idea simply in prudential terms, and it competes with other prudential, money-measured land-use alternatives. It is not framed as, or incorporated within, anything like the kind of frontal, moral challenge that is needed to reorient society. Just as bad, it doesn't call into question any material element or moral message of the capitalist market. It is, in truth, a way to slip a bit of ecological content into sustainability without really rocking the boat; without demanding, in a way that stands out, that the normative trinity of social ethics, political liberties, and market capitalism make room for a new, major player. It is deficient, that is, in the way that most environmental rhetoric is deficient—it speaks to people where they now are, with language consistent with the ways they think.

Environmental reform efforts need to draw attention to our overriding, shared need to succeed at our oldest task, which requires, among other labors, that we engage in the serious public work of discussing how we ought to live. That work needs to engage the hopes and ideas of many people. That being so, it won't help if environmental reformers claim up front to have all of the answers; if they show up with detailed blueprints for how people living in a landscape ought to behave. The danger here is, of course, especially high in the case of landscapes that are mostly in private hands.

What's needed instead is a full airing of the relevant line-drawing factors and an explanation of why they are pertinent, to our oldest task and as a matter of public, not merely private business. In particular settings, environmental groups might offer critiques of how landscapes are used. Land-planners might similarly come up with particular visions and proposals. But the main thrust of the reform effort cannot sensibly involve a green intelligentsia that goes around lecturing people how to live. Yes, stern language is needed. But it should take the form, not of directives on how to live, but rather as a call to show greater virtue and foresight, a call to be a good caretaker and, in time, an honored ancestor.

As early chapters explored, the first building block of good land use is respect for nature's ecological functioning, for the basic processes that keep the land fertile and maintain its primary productivity. That productivity is complexly linked, in ways still being studied, with biological diversity. Supporting this basic norm of ecological functioning would be the protection and restoration of wild places, managed under norms that minimize further human alteration; they can contribute to the health of larger landscapes while providing, as noted, independent value as well. Good land use would draw in and respect our best-considered moral judgments about the diffusion of moral value in nature, in other species and individual creatures. It would recognize and incorporate the limits on our knowledge of nature—our ignorance—and embrace an element of caution while drawing upon nature's embedded wisdom. It would include, as explored in chapter 6, the important social justice elements of good land use, often using them to insist that the good use of land in one location not occur at the expense of degradation elsewhere.

What is vital, as noted, is to get these ways of thinking into the public mainstream and get broader engagement with them, as matters of public morality and public business. More refined presentations might be assembled by way of illustrations, as ways of stimulating interest and productive thought. But detailed land-use plans would be, in most settings, premature.

Clean-Up Work

It ought to be clear that the reform effort put forth here is vastly different both from the current work of environmental organizations and from the various critiques of environmentalism circulating in recent years—critiques that are typically embedded in, and would merely perpetuate, our land-degrading culture. (The shrill call to broaden the base of support for environmentalism, for instance, is just a way of saying we need to work harder in ways that are accomplishing little and need to tailor rhetoric so that it is even more consistent with current modes of thinking.)

So strong and successful a generation and more ago, the US environmental reform effort, we must confess, has largely run aground. Once-effective ways of promoting environmental reform are no longer working. Even after a half-century of operation, the environmental movement as a whole has no real overall vision to present to citizens. Many citizens assume the movement is out to protect nature from people—to put as much of nature off limits as possible—and doesn't much care for places where people live and work. The assumption is quite wide off the mark, but it is easily entertained in the absence of clear evidence to the contrary. A half-century ago most Americans had no trouble naming the key leaders of the civil rights movement. Pressed today, most Americans could not name a single environmental leader, certainly not one whose fame does not arise for some other reason (movie roles, political office). For those who pay closer attention, there is little sense that the environmental movement takes ideas and rhetoric seriously except insofar as they can be used for short-term institutional gain. The rise of the political right, as often observed, was aided by close attention to ideas and rhetoric, carefully assessed and assembled through well-funded think tanks. While noting this impressive record of achievement, the environmental movement has made little effort to follow its lead.

Further insights into the current status and malaise of environmentalism can be teased out of the now-familiar photograph of a polar bear, precariously poised on fragmenting ice-sheet, displayed by environmental groups in the hope that viewers will send money or call a legislator. Perhaps the polar-bear image will indeed elicit a slighter stronger public response than alternative messages. But what messages are embedded in this heart-tugging photograph, particularly for viewers who see environmentalism as remote and elitist? A polar bear lives far away; environmentalism is thus about saving nature in distant places. The polar bear needs space to live away from people, so the movement protects nature from people. Polar bears meet no human needs nor do jobs depend on them. The photograph leaves human exceptionalism unquestioned; it promotes no ecological or communitarian values; it offers no challenge either to the market and market thinking or to the normative primacy of liberty and equality. To be sure the polar-bear photo invites concern for individual large mammals. But that sentiment, good so far as it goes, does little to advance the broader environmental agenda.

Just as the friendly polar-bear image needs reconsideration as a reform tactic, so too the environmental movement needs to reconsider its off-and-on interest in pushing a call for some sort of individual constitutional right to a sound environment. The idea here is that such a right, if embedded in constitutions, might stimulate major change. But would it? Starting around 1970 many individual states did amend their constitutions to proclaim such a constitutional right, often phrased as a right to a healthy or healthful environment. Calls to expand these rights are still heard, even as the rights, in the states that nominally recognize them, accomplish very little. Around the world, the rights-based strategy is even more pronounced.

The impulse behind this reform push is worthy enough, but there are grave limits and costs to it, so much so that the whole idea ought to be shelved, certainly in developed nations. One problem with such rights provisions is that they are simply too vague; they fail to give meaningful guidance on the line between use and abuse, particularly at the small spatial scales where legal disputes typically arise. They invite courts to do the line-drawing themselves, without guidance, a task for which courts are ill-suited by knowledge and temperament. Further, such provisions simply do not match the format of other constitutional rights. The individual-rights tradition regulates the link between citizen and state. It limits how and when government can interfere with people's lives; that is the standard format (the ban on slavery is much broader). In the case of environmental decline, however, much of it is brought about by private actors, not government. Even in the case of government-linked land abuse, the failings are often due to inaction rather than action. Finally, as phrased the rights tend to protect only against direct health threats to individuals living today, threats mostly (or entirely) due to pollution and contamination. They are not rights that protect land communities as such, ecological processes, other life forms, or future generations. At the international level, positive individual rights—to clean water, for instance—serve social justice and development goals and should be understood in that particular context. They are not sound means to protect the environment as such.

The bigger problem with the individual-rights approach to environmental protection is that it fits in much too easily with established ways of thinking and talking. It does not encourage new ways of perceiving and valuing nature as such; value continues to reside in humans only,

understood as autonomous, rights-bearing individuals. The value being exalted is the old, familiar, costly one: negative individual liberty.
Rights claims, as long known, can be divisive, and they collide with
opposing rights claims—in this instance with individual liberty and
private property. To argue about rights is to shift the fight to a cultural
playing field—human-centered, individualistic, present-focused, and
so on—that strongly favors opponents of environmental protection. As
a strategy, rights-promotion is badly flawed.

A Strategy

To move forward, the environmental reform effort needs a long-term
strategy for fundamental change. Long-term means decades, not
months or a few years. Fundamental change means in the major elements of modern culture, change in the playing field in which current
clashes unfold. Strategy means a carefully constructed plan that guides
pretty much all reform efforts and that screens out all program work—
and fundraising, membership appeals!—inconsistent with it. Communications needs to be more than an appendage of or support for the
conservation work of environmental organizations. It should become
the central component of that conservation work.

No single environmental organization can hope to push modern
culture in a new direction on its own. This hardly seems like an option. The option becomes plausible when many groups decide to work
together, when they decide not to glimmer as the thousand points of
light but to form a single, strong beam that draws attention. Concerted
action is essential, going well beyond, and indeed different in kind,
from the occasional joint campaigns that groups sometimes organize.
Groups would order their work and public messages to line up with
the overall reform strategy, a possibility perhaps hardly imaginable today but the only possibility that offers real hope. Coordinated action,
based on careful study, need not begin on all fronts at once; it could
start, for instance, with an orchestrated effort to put forth a new vision
of responsible private land ownership. But it needs to get started, breaking ground with clear thought on the shortcomings of current efforts
and with a full inquiry into the true root causes of land misuse.

Seeing and talking about nature—a new ontology. In the conservation
talks that Aldo Leopold gave to audiences in his final years, talks intended to push listeners in new directions, he routinely began by presenting a new understanding of nature and of the human place in

nature. The land was a community of life, of interconnected and interdependent biotic and biotic elements, a community that included people. Environmental reform today needs to press forward this same message, one that highlights the ways we are linked to nature and dependent on it. Sound messages could stress the reality and importance of the interconnections as such, the critical relations among the parts, the approach largely used by Pope Francis in his 2015 encyclical, *Laudato Si'*. It could sensibly employ more openly communitarian language, as Leopold did, particularly when presented as a normative vision rather than as a purely scientific claim. It is hardly conceivable that we might prune our misuses of nature without seeing nature in new ways, more ecologically grounded, which is to say pushing aside views of nature as a warehouse of physical market inputs.

A matter of shared morality. By steps we need to understand that our dealings with nature are matters of public concern, of legitimate public interest, and not merely matters for private resolution. Nature is our common home, where we all live. It is vital that we keep it healthy and productive; it is desirable that we keep it beautiful and pleasing, a place where human life and hopes can flourish. Natural conditions can be better or worse for us (for all life), which means that uses of nature raise basic normative issues, about right and wrong, wise and foolish, beautiful and ugly, just as human acts in nature are linked to individual virtue. To frame land use as a matter of shared concern, and in moral terms, is to remove it from the dominance of individual preference. It becomes a component of the common good, rightly talked about in that way. It becomes a matter for action by people wearing their citizen hats.

Varied language can be used to convey this component of the overall, culture-change strategy, chosen and shaped by setting and audience, including religious language. Moral claims can highlight the values of other species and individual creatures. They can surely give prominence to future generations and to the virtue of living now so that later generations flourish. The call can go out, in David Ehrenfeld's words, for us to become good ancestors. Rights rhetoric—species rights, animal rights—might fit in, but only if used with great care and blended with overriding messages that stress community, interdependence, and long-term human flourishing. Virtue-based language will likely work far better—the language of responsibility, good character, discipline, community membership, and clean living.

It hardly needs saying that morally charged messages are best framed so that they invite people to become better than they are, so that they look forward to healthier, more flourishing communities, and do not come across chiefly as condemnations—even as they are that, beneath the surface. Let listeners take the new morality and use it themselves to criticize current ways of living. It is important too to recognize openly that the community-level goal of healthy lands does create tension with highly valued norms of individual liberty. It would be wrong to hide this tension, and certainly wrong to deny it. It would be wrong too not to agree that tradeoffs will be needed. Yet it is possible also to point out that liberty comes in varied forms and that a long-time critical liberty is the positive, collective liberty of working with other community members to protect and enhance the community's home. Environmental reformers can be—they should be—very much in favor of liberty. What's needed on this issue (as on private property, considered below), is an alternative moral vision, in this instance one that emphasizes how options are broadened by collective action and how many environmental goals are only within reach when people work together.

As they put forward this message on land use and public morality, reformers should resist expected demands to translate their messages into specific policy changes. Inevitably, regularly, they will be pressed with the question: "So, what does this mean in terms of new laws or polices?" The question is pertinent, of course. But to answer it is to shift attention away from culture change, the key challenge. It is to turn the vital rhetoric about community, interconnection, public business, shared morality into mere background material, leading up to policy proposals, which will then (usually) be evaluated by audiences using old, familiar cultural frames. The reform effort needs to say focused on cultural change. If our culture shifts in good ways, our uses of nature will get better.

Drawing the line. A sound reform strategy, to reiterate, will emphasize the need to distinguish between the use and abuse of nature as a matter of public, moral business. For reasons noted, care must be used to avoid getting far ahead of audiences, particularly in playing the role of land-use expert and telling people how they ought to live. Three aims might be kept front and center.

First, it is vital to emphasize that we humans can legitimately change nature. All change is not abusive. The message may seem obvious, but the public is suspicious and no small number of environmental writers and activities implicitly return to unaltered nature as their benchmark of healthy lands. Yes, the environmental movement has been and must remain a voice for nature. But to succeed it needs to take on a more central role, encouraging audiences to bring competing interests and factors together to generate visions of good land use, ones that keep nature healthy and promote human flourishing.

Second, effective reform will push people to think broadly about the big questions: How should we be living in nature? What kinds of landscapes will best support the community of life, now and in the future? How might we make our natural homes more pleasing for us and our descendants? For most people these will be new questions, legitimate and worth considering only if audiences can see that the questions do raise issues for legitimate public action.

Third, having raised these issues, presenting them as normative ones—not matters of science or economics alone—the reform effort needs to stimulate thinking by interjecting the full range of relevant, line-drawing factors noted above. At the same time, and as important, it needs to insist that this line, once drawn, be used routinely to evaluate human uses of nature; to identify environmental problems. To return to the opening pages of chapter 2, the response to the question, "Why care about mussels in rivers?" needs to turn immediately to this use/abuse line: Mussels are dying because we are misusing our rivers, under standards of acceptable land and water use. Similarly, economic studies of alleged problems need to be challenged on the ground that they have failed (as they typically will fail) to take into account the relevant factors. The work here, of challenging flawed economic studies, will be vast, continuing, and exhausting.

In this line-drawing/cultural-change work, it will be vital to keep attention focused on the good that can come from this, vital to present healthy, flourishing landscapes as places we would want to inhabit. Accentuate the positive. What is less helpful—and perhaps positively hurtful—is to frame this public business chiefly in terms of limits. Yes, to be sure, we need to cut back in many ways. Yes, we need to embrace limits. But the point of respecting limits is to promote healthier lives and landscapes in the future. Framed in that way, behavior changes—limits included—are means to move in a good direction. Indeed, with new cultural frames, what we may be giving up are practices we come to see as wasteful or degrading. Good reform rhetoric would present matters in just that light.

Citizen action and good government. As noted, a sound environmental reform push needs to get people to act more often and forcefully in their roles as citizens rather than consumers, which is to say to support public policies to promote good land use. This part of the strategy will similarly be a long-term effort. It means confronting and revising the negative attitudes many people have toward government. Fortunately much of this animosity is aimed at the federal level and less at local and state levels. Land-use regulation, of course, has long been chiefly a local concern and wildlife law largely resides at the state level. In terms of changing attitudes, a useful approach might be to stress the ways well-run government could bring gains and then to ask: How can we improve government so that it delivers these desired goods? There is little reason to start off defending government as it is; the task is likely too much, though it is worthwhile to emphasize examples of government success. Better is to talk about what government could be and could do, and to encourage demands for improvement.

The reform push here of course also confronts the reality of governmental systems dominated by economic elites. The massive costs of this go well beyond environmental degradation. Environmental reformers need to lend a hand with efforts to challenge this grave and rising darkness. At the same time, it is important to recognize that this ominous trend is itself an expression and embodiment of grave flaws in modern culture, many of them the same flaws that help bring on land abuse and protect it from criticism. The takeover of government by market interests further shows the ways the market is pushing aside competing normative visions, playing upon America's love of individual liberty and its willingness to leave so many moral issues for resolution by individuals in private lives. On this point, the strong rights rhetoric of the political left is every bit as much to blame as the anti-government rhetoric of the political right. Indeed, many voices on the left seem even more insistent that moral choices be made by individuals as such, and that government leave people free to chart their own life courses. Hardly any rhetoric could serve better to open the door for moneyed interests to manipulate government to their advantage, as it has.

Major change in our sad political plight hardly seems possible without major cultural change. And the cultural change that would help environmental reform will also make inroads on the governmental front. Good government, if it comes, will emerge out of a heightened sense of the ways citizens are linked and interdependent; it will draw upon new recognitions of community and the ways we are the same, not different; it will gain strength when people talk more of the common good as a matter of public morality and of obligations of all people to cooperate as well as compete.

Owning nature. One of the more adamant obstacles that environmental reform has encountered is the institution of private property rights in nature and the tenacious attachment people have to it. In the common view, successfully pushed by the environmental opposition, environmental laws and regulations regularly infringe upon private property rights. Yes, Americans may want healthier environments, but they aren't willing to pursue them at the expense of undercutting this bedrock institution.

Without question one of the grave failings of the environmental movement has been its near silence on this issue, its failure to organize a coherent response other than to contend, day by day, that the costs of infringing property rights are offset by environmental gains.

The environmental reform effort will get hardly anywhere on private land-use issues, including (for instance) polluted runoff from fields, without engaging this issue. Private property well structured can help improve environmental outcomes and bring an array of other, well-known benefits. Environmental reform can be, and very much needs to be, openly in favor of private landownership. What it needs to put forth, coherently and consistently, is an alternative vision of what ownership means and why it exists. At present, the environmental movement seems to clash with private property so that property resides on one side of the clash. A far better way to frame the tradeoff is between responsible land ownership, on the environmental side, and irresponsible land ownership on the other. The dichotomy is simplistic, and far more needs to be said. But plenty of room exists, consistent with America's legal trajectory and culture, to formulate a vision of owning nature that expects owners to act in ways consistent with community needs and healthy lands. Under an alternative, community-sensitive vision of ownership, new land-use rules could make far better sense. With a new understanding of ownership it could make sense to limit landowners to activities on the moral side of the use/abuse line. Often the primary beneficiaries of good land use are other landowners themselves, which is why urban land-use rules are typically promoted more vigorously by landowners than anyone else. Cannot the environmental movement as a whole get together to study this topic and develop a shared stance?

On this issue, as on too many others, environmental opponents have skillfully slanted the playing field. In the typical property-rights story, big government interferes with the activities of a small, individual private landowner, restricting what she can do. Never does the anecdote draw other characters into the land-use drama, whether neighboring landowners, the local community, other life forms, or future generations. Never does it present the issue as one with moral overtones or as one in which caution makes sense. Certainly the landowner at issue is never a giant corporation that wields as much or more power as any government. In fact, land-use regulations almost always aim to accommodate conflicts in land- and resource-uses between competing property owners, so the typical conflict brings property owners to both sides of the table. A thoughtful, effective environmental stance on property ownership would also emphasize strongly the public's own property rights in water and wildlife and the implications of those property rights.

In talk about private ownership, the issue often turns to whether a particular landowner action is or is not harmful. Landowners admit, however begrudgingly, that they need to avoid causing harm. On this point, environmental reformers can insist that harm be measured using a well-considered normative standard of good land use and that it not somehow be treated as simply an issue of scientific fact or economic calculus. Certainly it is essential to steer far clear from the imposition of any scientific burden of proof. To the extent possible reformers ought to frame the issue as one of values and virtue; about the folly of charging ahead without knowing the consequences; of interfering with the health and beauty of other life. It is important, in short, not to let opponents frame the issue, as they usually have.

Forcefully, though very carefully, environmental reformers also ought to interject the idea that nature as a whole, all of it, is in some ways shared by all people and that public claims to it are overriding. The language here could be framed in terms of public ownership rights that stand above the private rights. The familiar image of stewardship pushes in this direction. One danger here, a big one, is that such language can readily draw the dreaded tag of socialism. On the other side, stewardship language can easily sound like a call simply for individuals to act responsibly as a matter of personal choice, free of communal insistence. In law-related writing, calls to expand the public trust ideal have encountered just such resistance, along with shrill claims that legal precedents are being misused. A better approach may be to leave untouched allodial landownership ideals (in which a private owner accounts to no one) while talking about the rightful role of lawmaking communities in demanding good behavior. Reform efforts can themselves make good use of the longstanding bans on land-use *harm* and on the normatively powerful claims that individual landowners should do their *fair shares* to remedy problems caused by landowners in combination.

Taming market capitalism. In a view ascending among environmental scholars, today's environmental ills are inextricably linked to modern capitalism and cannot be mitigated materially without fundamental changes in the market economy. The claim is sound, of course. But the tendency is to focus too much on specific economic practices and to discount or ignore the underlying cultural problems. As taken up in chapter 7 and again here, market capitalism is perhaps most pernicious because of the ways it frames and empowers the modern worldview. Within that worldview the institution makes decent sense and calls for structural change do not. A much different form of market capitalism can rise up only on a quite different cultural base. The challenge to market capitalism, therefore—and it very much needs challenge, just as critics say—can best push for cultural change.

Most of the strategic changes already included in this list will help weaken the grip of capitalism as we know it:

· A new vision of nature as an ecologically complex, interconnected whole calls into question the market's much different view. Nature is not mere raw materials, it is our common home, infused with various moral values.

· A stronger vision of good land use can weaken the idea that market suppliers are free to act as they see fit to cut costs and make money. The current market view will also weaken as individual negative liberty takes up less space on the stage of public morality.

· The market currently rests on a particular understanding of private landownership and of the limited role of government in restricting land use options; here again, cultural changes would cast doubt on many market activities.

· In the market view, individuals are cast as producers and consumers. The market's role contracts when and as they act instead as citizens.

· As Herman Daly and others have long urged, the nation could benefit from new measures of overall national well-being, competing with the now-powerful GDP, particularly measures linked to environmental health.

· Another sound reform often mentioned is to challenge the relatively new dominance of shareholder capitalism as the guiding ideal for corporate management. In the now-prevailing view, corporations exist solely to make money for shareholders. In the older, long-dominant view, corporations existed instead to promote the interests of multiple stakeholders, including employees and local communities. On this issue, as on several others, environmental reformers can join forces with other causes similarly harmed by this profoundly hurtful shift.

· Finally, the power of market capitalism would diminish if states and local governments had a freer hand to set higher environmental standards for interstate and global businesses operating within their bounds. Many jurisdictions of course would not exercise this power, but some certainly would. As a legal matter the

Constitution's Dormant Commerce Clause restricts state and local powers too much. Environmental reformers could join with others to push against it.

These actions and others could help chip away at the ills of contemporary market capitalism. But it is essential, to reiterate, to keep the cultural issues front and center. The market builds on a certain set of values and understandings. To disturb them is to weaken its foundation.

One long-time call for changes in market capitalism takes the form of a demand to shift to a steady-state economy, or in some versions to promote de-growth. The ideas are sound insofar as they recognize that patterns of altering and consuming nature need to change significantly, and the chosen language pushes this reality hard. Again, though, one must question whether such language does not unhelpfully accentuate the negative. It puts emphasis on what is being given up, not what is being gained. It casts accusations in particular at people simply trying to make a go of things in the economy as it now exists, and features a call to restrict current opportunities at a time when, for most people, they are meager enough already. Far better rhetoric and packaging can be used to promote the same changes in culture and economic practices. In truth, it is language like this—so negative, so threatening—that goes far toward explaining why environmental reform struggles while economic liberty platforms gain strength. Desired changes are one thing; the best rhetoric to promote them might well be far different. Anti-environmental forces don't frame their stances in terms of opposition to environmental laws. Instead they are out to protect and promote good things such as private property, individual liberty, and economic opportunity. Cannot environmental reformers learn from their success?

As environmental advocates look out at modern society, they see entrenched cultural opposition, and rightly so. Vast audiences are immune to scientific information. Economic studies go ignored, or are countered by competing studies that are so flimsy as to lack any credibility. The seeming idiocy of the national Republican Party in denying the human role in climate change, if not climate change itself, is merely the most vivid and infuriating example.

So what language might be used to reach out to these fellow citizens, who beneath the surface are often just as worried about our current predicament and just as desirous of change that leads to a better future? The answer: rhetoric that starts with their moral values and builds on them:

- Most such citizens value their communities, their home lands and cultures, and are motivated to defend them. Yes they embrace liberty, but quite often in just the way America's revolutionaries did, as language of defense. Green messages to them need to be framed in terms of their community and the common good.
- In the conservative mind virtue and good character remain important, more important, it must be confessed, than it is to many progressive liberals. Good citizens, landowners in particular, are people who act responsibly, who are good community members. Virtuous people do not harm others; virtuous people care about generations to come. What does it mean to be a person of honor and dignity in a world that is crowded and polluted?
- Virtue and community membership come together in the normative ideal of doing one's fair share, of helping remedy shared problems. When the common good includes healthy lands, those who degrade lands and waters are acting immorally. Indeed, they are stealing from the common fund.
- In the conservative moral view, fertility and productive are good, waste is wrongful, and contamination is repulsive.
- Yes, liberty is good, but taken too far it is license, and liberty is also exercised when local people got together to improve their communities, drawing on the old barn-raising tradition of neighborliness.
- Private property is a sound institution and it works well when owners show good character. Responsible ownership is all-American; irresponsible ownership needs to be challenged. A responsible owner does not act in ways that harm other owners, the community, or future generations.

The Next 250 Years

The normative worldview of the United States, a worldview now greatly influential globally, came together in the decade leading up to the American Revolution of 1776. Moral rhetoric then did feature strands of communitarian, civic-republican thought and organic social hierarchies continued thereafter to carry power. But a new vision was coming together, more individualistic, more liberty-based, more insistent on equality. As the decades unfolded, it made sense to embrace normative orders that elevated individuals and honored their rights, morality that philosophers would term deontological. It made sense similarly to talk about utility-enhancement as an individual and shared goal and to measure utility in terms of how well it responded to individual wants, the morality termed utilitarianism or consequentialism. Christian compassion motivated many reform efforts, even those clothed in the garb of individual rights, but compassion and rights were not always good bedfellows. We see that clearly today in the furor of government-subsidized health care for those without it, a compassionate move some say, a violation of liberty others say, particularly liberty of the eat-only-what-you-kill variety.

It was once thought, for decades really, that environmental reform built upon and continued earlier civil rights campaigns. Just as moral rights and equality expanded for racial minorities, women, and the disabled, so too the moral circle could broaden to include other life forms. Environmental protection was the next step. In his useful review of environmental ethics, *The Rights of Nature*, historian Roderick Nash told the story in just this way. The time had come for nature to have rights of its own. Nash and others often quoted the vivid, opening paragraphs of Aldo Leopold's famous essay, "The Land Ethic," from 1949. In them, Leopold spoke of the expansion of moral concern since the ancient times of Odysseus, who freely hanged slaves who had misbehaved. Moral concern had expanded since then, Leopold observed. It was time, he said, for it to take the next step and include the land community as such.

Not enough attention was paid to the fact that Leopold's planned expansion of moral value did not form a clear sequence with earlier expansions, or with the kinds of expansions that later animal-welfare writers had in mind. Leopold jumped from present-day, living humans to a vast, interconnected community of life. He did not propose the more logical step of adding individual, sentient mammals to the moral circle, leaving intact the emphasis on individual creatures understood as autonomous beings. In fact, Leopold's proposal was for a vast overhaul in moral thought, building upon a similarly vast overhaul in ways of understanding humans and their natural homes. Leopold's essay drew widespread support, but even as it did academic philosophers, more alert to his shift, typically scoffed at the mere idea that a community as such—even if it really existed as an identifiable thing—could possess moral value.

Nash's claim, consistent with then-prevailing thought, assumed that environmental progress could move ahead by drawing upon the moral materials that had served so well in the hands of other reformers. It could move ahead by talking about the rights of nature, individual constitutional rights to a healthful environment, and—as Rachel Carson did—how pollution and contamination violated the rights of people harmed by it unless they gave informed consent.

Reformers who embraced rights rhetoric faced problems in the late nineteenth century when big business latched on to the rhetoric of liberty and private property to resist much governmental regulation. But progressives regrouped, drew more on Christian social ethics, and pushed on ahead in the early twentieth century and, after World War II, with civil rights campaigns. The morality, the language of political rights, seemed to work still, even as conservative groups saw more and more ways to use it themselves.

What the past forty years ought to make clear is that the civil rights–type moral frame no longer works. Yes, the marriage equality campaign has used it well, so too campaigns for the physically limited. But the examples are narrow. More glaring are the failures, not just ecological degradation but rising economic injustice and the decline of anything resembling real popular sovereignty. And the trends are ominous; in David Orr's words, we are in failure mode, not so much rearranging deck chairs on the *Titanic* but, in his better, borrowed metaphor, walking north in a southbound train. Looking ahead, rights-based claims seem likely to gain traction only when they do not materially interfere with big business carrying on as usual (as marriage equality and disability gains do not). Environmental reforms do benefit many businesses, new ones especially. But that's not nearly enough. They collide with business as usual, in very big ways.

Without too much simplicity we might say that our current moral order has now been in place in the United States for 250 years. We are overdue for a major change of course. For the next 250 years, maybe more, we need to pursue a different path. The trajectory turn that began in the 1760s took time to get going. But it gained momentum and strength for the next century, through the slavery-ending Civil War. Looking around, there is evidence already that a new turn has begun. A strong environmental reform effort, well considered and orchestrated, can give it ideas and values, direction, and considerable power.

Acknowledgments

Because this book draws together work I've done over more than thirty years it is hardly possible to take note of the many people who have helped along the way, some who did so knowingly, others who did so from afar or in ways they did not recognize. Much of the first draft was written during the fall of 2013 while I enjoyed the hospitality of the Law Faculty of the University of Cambridge and the special friendship of Kevin Gray of Trinity College, to whom I am much indebted. I benefited too in the early stages from an extended stay as Visiting Fellow at the Stellenbosch Institute for Advanced Study in South Africa, an exceptional place to witness and study the ways that peoples, lands, and culture intermingle. Otherwise the work was largely done at my home institution, the College of Law of the University of Illinois at Urbana-Champaign, which has been warmly supportive over the years in so many ways, not the least by providing a setting that exemplifies, encourages, and honors intellectual achievement. My home has been a good one also because, with intensively used farm monocultures all around, one never forgets how hard and thoroughly modern culture and the capitalist market have pressed against and radically simplified the natural order.

Many of the ideas set forth here were clarified and presented to arrays of undergraduate and graduate students and members of the public in a campus-wide series on environmental issues held each spring at UIUC for the past seven years. I'm grateful for the chance to play a key role in the series from the start and grateful particularly for the many students, faculty colleagues, and other speakers who in varied ways help hone my thinking. Professor Robert McKim has been with me from the start and aided considerably along the way, not just by answering questions and commenting on pages but showing me the exceptional values of a first-rate philosophic mind. I am similarly indebted, for valuable discussions and comments, to philosophers J. Michael Scoville, Heidi Hurd, Alexis Dyschkant, and Rob Kar. Rob Kanter and Fran Harty have served as role models in their embrace of local nature and their abilities to awaken audiences to it. They have helped too in reminding me that biological riches can still be found and celebrated, even extensively simplified landscapes. During his student years at Illinois, Tom Rice was particularly helpful and inspiring. While a student years ago and consistently since them, Todd Wildermuth has been an especially useful sounding board and source of ideas. Nick Fregeau and Zviko Chadambuka were kind enough to read the manuscript in full and comment on it; I thank them for doing so.

In all matters related to the environmental movement and its successes and failures—and for answer to countless questions about science and technology—I have turned for many years to Clark Bullard, whose knowledge extends far beyond any conception of his home discipline in mechanical engineering. I have learned much, too, from my many colleagues over the years at the Prairie Rivers Network of Champaign—an exemplary organization—and from fellow parishioners and senior pastors of University Place Christian Church, Champaign, particularly, during the writing of this work, Pastor Kristine Light. At the University of Chicago Press, editor Christy Henry has been firmly supportive of my work, on this book and another, from the very moment I mentioned them to her. I thank her much for it.

As with any scholar whose works draws heavily upon the writings of others my greatest debts are to scholars whose books I have read, and in some cases underlined and reread over and over. The major books, and the authors of them, are set forth in the selected bibliography. Particularly as my work on this book neared its end I became more aware than ever that the cultural challenges we face today have been with us in varied forms for generations, and that much of the most penetrating commentary on them is not recent. I thus found myself returning repeatedly to volumes by Joseph Wood Krutch, Aldo Leopold, and Lewis Mumford, among others. As for intellectual light and inspiration and, more particularly now, for commenting on this manuscript, I especially thank historian Donald Worster.

Notes

INTRODUCTION

1. Aldo Leopold, "Engineering and Conservation," in *The River of the Mother of God and Other Essays by Aldo Leopold*, ed. Susan L. Flader and J. Baird Callicott (Madison: University of Wisconsin Press, 1991), 254.
2. Louis Menand, *The Metaphysical Club: A Story of Ideas in America* (New York: Farrar, Straus, Giroux, 2001), x.
3. Daniel T. Rodgers, *Age of Fracture* (Cambridge: Harvard University Press, 2011), 41.
4. Edward O. Wilson, *The Social Conquest of Nature* (New York: W. W. Norton, 2012), 243.
5. Richard M. Weaver, *Ideas Have Consequences* (Chic`ago: University of Chicago Press, 1948), 41.

CHAPTER 4

1. Abraham Lincoln, "Address" at Sanitary Fair, Baltimore, Maryland (April 18, 1864), in *Collected Works of Abraham Lincoln*, vol. 7, ed. Roy P. Basler (New Brunswick, NJ: Rutgers University Press, 1953), 302.

CHAPTER 5

1. *Commonwealth v. Alger*, 7 Cush. 53, 84–85 (Mass. 1851).
2. *Charles River Bridge v. Warren Bridge*, 36 U.S. (11 Pet.) 420, 548 (1837)
3. William Ophuls, *Plato's Revenge: Politics in the Age of Ecology* (Cambridge, MA: MIT Press, 2011), 34–35.

CHAPTER 6

1.   C. S. Lewis, *That Hideous Strength* (New York: Macmillan, 1946), 178.
2.   Jean-Jacques Rousseau, *Discourse on the Origin and Foundations of Inequality among Men* (New York: Bedford/St. Martins, 2011), 70.
3.   Quoted in Stanley N. Katz, "Thomas Jefferson and the Right to Property in Revolutionary America," *Journal of Law and Economics* 19 (1976): 480 (from letter dated October 28, 1785).
4.   Quoted in John F. Hart, "Land Use Law in the Early Republic and the Original Meaning of the Takings Clause," *Northwestern Law Rev.* 94 (2001): 1126.
5.   Thomas More, *Utopia* (New Haven, CT: Yale University Press, 1964), 76.
6.   Thomas Paine, "Agrarian Justice," in *The Thomas Paine Reader*, ed. Michael Foot and Isaac Kramnick (New York: Penguin Books, 1987), 476–78.

CHAPTER 7

1.   Wendell Berry, "The Whole Horse," in *The Art of the Commonplace: The Agrarian Essays of Wendell Berry* , ed. Norman Wirzba (Washington, DC: Counterpoint Press, 2002), 236.
2.   Ibid.
3.   Martin Gilens and Benjamin I. Page, "Testing Theories of American Politics: Elites, Interest Groups, and Average Citizens," *Perspective on Politics* (Fall 2014).

CHAPTER 8

1.   Rogers, *Age of Fracture*, 3, 6.
2.   Ibid., 19–20, 35–36.

Selected Bibliography

Abbey, Edward. *Desert Solitaire: A Season in the Wilderness*. New York: McGraw-Hill, 1968.

Abrams, Richard M. *America Transformed: Sixty Years of Revolutionary Change, 1941–2001*. New York: Cambridge University Press, 2006.

Alperovitz, Gar. *America beyond Capitalism: Reclaiming Our Wealth, Our Liberty, and Our Democracy*. Hoboken, NJ: John Wiley and Sons, 2005.

Barber, Benjamin R. *Strong Democracy: Participatory Politics for a New Age*. 20th anniv. ed. Berkeley: University of California Press, 2003.

Becker, Carl L. *The Heavenly City of the Eighteenth-Century Philosophers*. New Haven, CT: Yale University Press, 1932.

Bell, Daniel. *The Cultural Contradictions of Capitalism*. 20th anniv. ed. New York: Basic Books, 1996.

Bellah, Robert N., et al. *Habits of the Heart: Individualism and Commitment in American Life*. Updated ed. Berkeley: University of California Press, 1996.

Berman, Morris. *The Twilight of American Culture*. New York: W. W. Norton, 2000.

Berry, Wendell. *The Unsettling of America: Culture and Agriculture*. San Francisco: Sierra Club Books, 1977.

Boehm, Christopher. *Moral Origins: The Evolution of Virtue, Altruism, and Shame*. New York: Basic Books, 2012.

Bollier, David. *Silent Theft: The Private Plunder of Our Common Wealth*. London: Routledge, 2003.

Brinkley, Alan. *Liberalism and Its Discontents*. Cambridge, MA: Harvard University Press, 1998.

Callicott, J. Baird. *Beyond the Land Ethic: More Essays in Environmental Philosophy*. Albany: SUNY Press, 1999.

————. *In Defense of the Land Ethic: Essays in Environmental Philosophy.* Albany: SUNY Press, 1989.

Chang, Ha-Joon. *23 Things They Don't Tell You about Capitalism.* New York: Bloomsbury, 2011.

Crunden, Robert M., ed. *The Superfluous Men: Conservative Critics of American Culture, 1900–1945.* Wilmington, DE: ISI Books, 1999.

Daly, Herman E., and John B. Cobb Jr. *For the Common Good: Redirecting the Economy toward Community, the Environment, and a Sustainable Future.* Boston: Beacon Press, 1989.

Delbanco, Andrew. *The Real American Dream: A Meditation on Hope.* Cambridge, MA: Harvard University Press 1999.

Diamond, Jared. *Guns, Germs, and Steel: The Fates of Human Societies.* New York: W. W. Norton, 1997.

Eckersley, Robyn. *Environmentalism and Political Theory: Toward an Ecocentric Approach.* Albany, NY: SUNY Press, 1992.

————. *The Green State: Rethinking Democracy and Sovereignty.* Cambridge, MA: MIT Press, 2004.

Ehrenfeld, David. *The Arrogance of Humanism.* New York: Oxford University Press, 1978.

Etzioni, Amitai. *The New Golden Rule: Community and Morality in a Democratic Society.* New York: Basic Books, 1996.

Evernden, Neil. *The Natural Alien: Humankind and Environment.* Toronto: University of Toronto Press, 1985.

Fawcett, Edmund. *Liberalism: The Life of an Idea.* Princeton, NJ: Princeton University Press, 2014.

Flader, Susan L., and J. Baird Callicott, eds. *The River of the Mother of God and Other Essays by Aldo Leopold.* Madison: University of Wisconsin Press, 1991.

Foster, John Bellamy. *The Ecological Revolution: Making Peace with the Planet.* New York: Monthly Review Press, 2009.

Fowler, Robert Booth. *The Greening of Protestant Thought.* Chapel Hill: University of North Carolina Press, 1995.

Frank, Robert H. *The Darwin Economy: Liberty, Competition, and the Common Good.* Princeton, NJ: Princeton University Press, 2011.

Genovese, Eugene D. *The Southern Tradition: The Achievement and Limitations of American Conservatism.* Cambridge, MA: Harvard University Press, 1994.

Gray, John. *Enlightenment's Wake.* London: Routledge, 1995.

Greenblatt, Stephen. *Swerve: How the Modern World Became Modern.* New York: W. W. Norton, 2011.

Hacker, Jacob S., and Paul Pierson. *Winner-Take-All Politics: How Washington Made the Rich Richer—and Turned Its Back on the Middle Class.* New York: Simon and Schuster, 2010.

Harari, Yuval Noah. *Sapiens: A Brief History of Humankind.* London: Harvill Secker, 2014.

Hays, Samuel P. *Beauty, Health, and Permanence: Environmental Politics in the United States, 1955–1985.* Cambridge: Cambridge University Press, 1987.

Hind, Dan. *The Return of the Public.* London: Verso, 2010.

Hobsbawm, Eric. *Industry and Empire: The Birth of the Industrial Revolution.* London: Penguin Group, 1999.

Horkheimer, Max, and Theodor W. Adorno. *Dialectic of Enlightenment: Philosophical Fragments.* Translated by E. Jephcott. Stanford, CA: Stanford University Press, 2002.

Israel, Jonathan. *A Revolution of the Mind: Radical Enlightenment and the Intellectual Origins of Modern Democracy.* Princeton, NJ: Princeton University Press, 2010.

Jackson, Wes. *New Roots for Agriculture.* Rev. ed. Lincoln: University of Nebraska Press, 1985.

Johnson, Mark. *Morality for Humans: Ethical Understanding from the Perspective of Cognitive Science.* Chicago: University of Chicago Press, 2014.

Joyce, Richard. *The Evolution of Morality.* Cambridge, MA: MIT Press, 2006.

Kammen, Michael. *Spheres of Liberty: Changing Conceptions of Liberty in American Culture.* Madison: University of Wisconsin Press 1986.

Krutch, Joseph Wood. *The Measure of Man: On Freedom, Human Values, Survival and the Modern Temper.* 1954; Gloucester, MA: Peter Smith, 1978.

———. *The Modern Temper: A Study and a Confession.* New York: Harcourt, Brace and Co., 1929.

Leopold, Aldo. *A Sand County Almanac and Sketches Here and There.* New York: Oxford University Press, 1949.

Lieberman, Matthew D. *Social: Why Our Brains Are Wired to Connect.* New York: Broadway Books, 2013.

MacIntyre, Alasdair. *After Virtue.* 2nd ed. Notre Dame, IN: University of Notre Dame Press, 1984.

Magdoff, Fred, and John Bellamy Foster. *What Every Environmentalist Needs to Know about Capitalism.* New York: Monthly Review Press, 2011.

Martin, Calvin Luther. *In the Spirit of the Earth: Rethinking History and Time.* Baltimore, MD: Johns Hopkins University Press, 1992.

McNeill, J. R. *Something New Under the Sun: An Environmental History of the Twentieth-Century World.* New York: W. W. Norton, 2000.

Menand, Louis. *The Metaphysical Club: A Story of Ideas in America.* New York: Farrar, Straus, Giroux, 2001.

Mill, John Stuart. *On Liberty.* 1859; Indianapolis: Bobbs-Merrill Co., 1956.

———. *Utilitarianism.* 1861; New York: Barnes and Noble, 2005.

Mumford, Lewis. *Technics and Human Development.* New York: Harcourt Brace Jovanovich, 1967.

Newton, Julianne Lutz. *Aldo Leopold's Odyssey: Rediscovering the Author of A Sand County Almanac.* Washington, DC: Island Press, 2006.

Nichols, John, and Robert W. McChesney. *Dollarocracy: How the Money and Media Election Complex Is Destroying America*. New York: Nation Books, 2013.

Noble, David W. *Debating the End of History: The Marketplace, Utopia, and the Fragmentation of Intellectual Life*. Minneapolis: University of Minnesota Press, 2012.

Norton, Bryan G. *Searching for Sustainability: Interdisciplinary Essays in the Philosophy of Conservation Biology*. Cambridge: Cambridge University Press, 2003.

Northcott, Michael S. *A Moral Climate: The Ethics of Global Warming*. London: Darton, Longman and Todd, 2007.

Ophuls, William. *Plato's Revenge: Politics in the Age of Ecology*. Cambridge, MA: MIT Press, 2011.

———. *Requiem for Modern Politics: The Tragedy of the Enlightenment and the Challenge of the New Millennium*. Boulder, CO: Westview Press, 1997.

Orr, David W. *Hope Is an Imperative: The Essential David Orr*. Washington, DC: Island Press, 2011.

Osborne, Roger. *Civilization: A New History of the Western World*. New York: Pegasus Books, 2006.

Pagden, Anthony. *The Enlightenment and Why It Still Matters*. New York: Random House, 2013.

Pells, Richard H. *Radical Visions and American Dreams: Culture and Social Thought in the Depression Years*. New York: Harper and Row, 1973.

Polanyi, Karl. *The Great Transformation: The Political and Economic Origins of Our Time*. 1944; Boston: Beacon Press, 1957.

Ponting, Clive. *A New Green History of the World: The Environment and the Collapse of Great Civilizations*. Rev. ed. New York: Penguin Books, 2007.

Richard, Carl J. *The Battle for the American Mind: A Brief History of a Nation's Thought*. Lanham, MD: Rowman and Littlefield, 2004.

Robinson, Marilynne. *Absence of Mind: The Dispelling of Inwardness from the Modern Myth of the Self*. New Haven, CT: Yale University Press, 2010.

Rodgers, Daniel T. *Age of Fracture*. Cambridge, MA: Harvard University Press, 2011.

———. *Contested Truths: Keywords in American Politics since Independence*. New York: Basic Books, 1987.

Rogers, Raymond A. *Nature and the Crisis of Modernity: A Critique of Contemporary Discourse on Managing the Earth*. Montreal: Black Rose Books, 1994.

Sandel, Michael J. *Democracy's Discontent: America in Search of a Public Philosophy*. Cambridge, MA: Harvard University Press, 1996.

Saul, John Ralston. *The Unconscious Civilization*. New York: Free Press, 1995.

———. *Voltaire's Bastards: The Dictatorship of Reason in the West*. New York: Free Press, 1992.

Schlatter, Richard. *Private Property: The History of an Idea*. London: George Allen and Unwin, 1951.

Scoville, J. Michael. "Environmental Values, Animals, and the Ethical Life." PhD diss., University of Illinois at Urbana-Champaign, 2011.

Sen, Amartya. *On Ethics and Economics*. Oxford: Basil Blackwell, 1987.

Shapiro, Ian. *The Moral Foundations of Politics*. New Haven, CT: Yale University Press, 2003.

Siedentop, Larry. *Inventing the Individual: The Origins of Western Liberalism*. Cambridge, MA: Harvard University Press, 2014.

Singer, Peter. *The Expanding Circle: Ethics, Evolution, and Moral Progress*. Rev. ed. Princeton, NJ: Princeton University Press, 2011.

Steinberg, Ted. *Down to Earth: Nature's Role in American History*. 2nd ed. New York: Oxford University Press, 2009.

Szasz, Andrew. *Shopping Our Way to Safety: How We Changed from Protecting the Environment to Protecting Ourselves*. Minneapolis: University of Minnesota Press, 2007.

Tarnas, Richard. *The Passion of the Western Mind: Understanding the Ideas that Have Shaped Our World View*. New York: Crown Publishers, 1991.

Tawney, R. H. *The Acquisitive Society*. 1921; Brighton: Wheatsheaf Books, 1982).

Thomas, Keith. *Man and the Natural World: A History of the Modern Sensibility*. New York: Pantheon Books, 1983.

Tomasello, Michael. *A Natural History of Human Thinking*. Cambridge, MA: Harvard University Press, 2014.

Twelve Southerners. *I'll Take My Stand: The South and the Agrarian Tradition*. 1930; Baton Rouge: LSU Press, 1977.

Walzer, Michael. *Spheres of Justice: A Defense of Pluralism and Equality*. New York: Basic Books, 1983.

Watts, Alan W. *Nature, Man and Woman*. New York: Pantheon Books, 1958.

Weaver, Richard M. *Ideas Have Consequences*. Chicago: University of Chicago Press, 1948.

Webb, Sidney, and Beatrice Webb. *The Decay of Capitalist Civilization*. New York: Harcourt, Brace, 1923.

Weber, Max. *The Protestant Ethic and the Spirit of Capitalism*. Rev. ed. 1920; New York: Oxford University Press, 2011.

Weston, Burns H., and David Bollier. *Green Governance: Ecological Survival, Human Rights, and the Law of the Commons*. New York: Cambridge University Press, 2013.

White, Morton. *The Philosophy of the American Revolution*. New York: Oxford University Press, 1978.

Williams, Chris. *Ecology and Socialism: Solutions to Capitalist Ecological Crisis*. Chicago: Haymarket Books, 2010.

Wilson, Edward O. *The Social Conquest of Earth*. New York: Liveright Publishing, 2012.

Wirzba, Norman, ed. *The Art of the Commonplace: The Agrarian Essays of Wendell Berry*. Washington, DC: Counterpoint Press, 2002.

Wood, Mary Christina. *Nature's Trust: Environmental Law for a New Ecological Age*. New York: Cambridge University Press, 2014.

Worster, Donald. *Dust Bowl: The Southern Plains in the 1930s*. New York: Oxford University Press, 1979.

———. *Nature's Economy: A History of Ecological Ideas*. 2nd ed. Cambridge: Cambridge University Press, 1994.

———. *Shrinking the Earth: The Rise and Decline of American Abundance*. New York: Oxford University Press, 2016.

———. *The Wealth of Nature: Environmental History and the Ecological Imagination*. New York: Oxford University Press, 1993.

Index